CROYDON: TRAM TO TRAMLINK

ALAN YEARSLEY

© 2023 Platform 5 Publishing Ltd. All rights reserved. No part of this publication may be reproduced or transmitted in any form or by any means electronic, mechanical, photocopying, recording or otherwise, without prior permission of the publisher.

Published by Platform 5 Publishing Ltd, 52 Broadfield Road, Sheffield, S8 0XJ. England.

Printed in England by The Amadeus Press, Cleckheaton, West Yorkshire

ISBN: 978 1 915984 05 0

Front cover (top left): Covered top eight-wheel car 22E on Thornton Heath–Croydon route 42 sometime between 1933 and 1936–37 when this class of tram was withdrawn. These trams were built as open toppers by Milnes in 1902 with their top covers being fitted in 1928 and the E suffix added to their numbers by the LPTB in 1933. *M.J. O'Connor/National Tramway Museum collection*

Front cover (top right): Variotram 2555 passes the eastbound platform at Lebanon Road with a Wimbledon service on 17 April 2022 as a passenger validates her card or device on the card reader. This stop has staggered platforms, with the westbound platform being just behind the camera. *Robert Pritchard*

Front cover (main photo): CR4000 car 2541 is about to cross the A212 road between Coombe Lane and Gravel Hill with a New Addington service on 17 April 2022. *Robert Pritchard*

Back cover: Variotram 2554 received a special Croydon Council "Love Croydon" promotional livery in June 2012 shortly after entering service, and still carries this colour scheme at the time of writing. On 17 April 2022 it is seen at East Croydon with an Elmers End service. *Robert Pritchard*

CROYDON: TRAM TO TRAMLINK

CONTENTS

Introduction 3
1. **Croydon's First Trams: 72 Years of Service** 4
   Horse to Steam and Gas 4
   Gas to Electric 5
   Decline, Closure, Bus & Trolleybus Replacement 6
   Rapid Transit Tramway Proposal 7
   The Two Former Heavy Rail Routes 7
   Wimbledon–West Croydon 8
   Elmers End–Addiscombe/Sanderstead 8
   The Never Realised Southern Heights Light Railway 12
   Preserved London Trams and Trolleybuses 13
2. **The Tramlink Project** 17
   Bus or Light Rail? 18
   Construction Starts 18
   Tramlink Opens 20
   Effects on Bus Services 21
   Tramlink Feeder Bus Routes 21
   Making a Difference? 22
   Ownership Changes 22
3. **London Trams Route by Route** 24
   London Trams map 25
   New Addington 26
   Beckenham Junction/Elmers End 33
   Central Croydon Loop 38
   Wimbledon 43
4. **The Tram Fleet** 53
   The CR4000s 53
   CR4000 Modifications 56
   CR4000 Liveries 57
   The Stadler Variotrams 61
   Variotram Liveries 61
5. **The Depot and Infrastructure** 65
   Track Renewals 70
   Signalling and Signs 74
   The Power System and Overhead Lines 74
   The Depot and Control Centre 74
   Therapia Lane Depot Goes Greener 78
6. **Services** 79
   Ticketing and Fares 85
   Ridership Figures 88
   Accidents 88
   The 2016 Sandilands Accident 90
7. **Future Developments** 93
   The Cross-River and West London Tram Schemes 94
   Abbreviations Used in this Book 95
   The Jubilee Line Extension: What Might Have Been 96
   Further Reading 96

INTRODUCTION

It is now 23 years since the opening of the London Trams (originally known as Croydon Tramlink) light rail system. Since then the network has established itself as an invaluable part of the area's public transport offering, even if a number of extensions have been proposed but have so far not materialised. Nonetheless, the system has helped to regenerate some of the areas that it serves and created a number of new journey opportunities by providing through services between previously separate parts of the local rail network.

This book provides an ultimate present-day guide to London Trams, the history and background to the Tramlink project, and its operations and vehicle fleets. Although the main purpose of this book is to cover the present-day London Trams network, we have also included historical material on Croydon's original tram and trolleybus services along with the former heavy rail lines on whose trackbeds the trams now operate.

This is the fourth in our series of books covering the UK tram and light rail systems following those on Sheffield Supertram, Manchester Metrolink and Nottingham Express Transit: it is planned to produce a similar publication on each network over the coming years. As they are published, these will be advertised in our magazines Today's Railways UK and Today's Railways Europe, on the Platform 5 Publishing website, and in our mail order catalogues and circulars.

We hope you will find this book an interesting and useful source of information. It draws on extensive research trips to Croydon over the years as well as original material from the launch of Tramlink, earlier Platform 5 publications including our own UK Metro & Light Rail Systems handbook and back copies of **entrain** and **Today's Railways UK** magazines, various editions of the Light Rail Review series published in the 1990s, and other books on Croydon's original and present-day trams such as Croydon Tramways by Robert J Harley, published by Capital Transport Publishing, and Croydon Tramlink by Gareth David, published by Pen & Sword.

I have made every effort to ensure that all information is correct at the time of going to press but cannot be held responsible for any errors or omissions. Nonetheless any corrections or suggestions for improvements for any future editions would be most gratefully received. Any comments on this publication should be sent by email to **updates@platform5.com** or by post to the Platform 5 address on the title page.

ACKNOWLEDGEMENTS

Most of the photographs used in this book have been drawn from many dozens of personal trips by the author and other Platform 5 staff and correspondents to Croydon to both ride on and photograph the tramway since it opened in 2000, and we are pleased to have been able to illustrate all 36 of the existing fleet of Croydon trams at least once within the pages of this book! We are also indebted to our correspondents Nick Comfort and Keith Fender for their proof checking, for all who have supplied us with present day and archive photos, and to the staff of TfL press office, especially David Edwards.

We also extend our thanks to the National Tramway Museum and the British Trolleybus Society for access to their archive photographic collections.

If readers have any photos that they would like to be considered for our forthcoming light rail system guidebooks (particularly any illustrating the early years of operation) please do get in touch using the email **pictures@platform5.com**.

UPDATES

Any major developments with London Trams, and the country's other light rail and tram systems, can be found in the magazine **Today's Railways UK**. This is available at all good newsagents or on post-free subscription. Please see the inside covers of this book for further details.

Alan Yearsley. July 2023.

Above: Variotram 2563 calls at Wandle Park with a Wimbledon–Elmers End service on 18 April 2022. *Robert Pritchard*

CROYDON: **TRAM TO TRAMLINK**

CHAPTER 1:

CROYDON'S FIRST TRAMS:
72 YEARS OF SERVICE

Above: This photo, taken from an old print, shows a horse tram crossing the station bridge at West Croydon in about 1900.

In common with many other areas of London, Croydon was served by an extensive tramway network in the late 19th and early 20th century. Croydon's original tramway had its origins in the Croydon Tramways Company (CTCo), which ran the first horse tram service in the town on 6 October 1879.

The initial system was in two halves, consisting of routes from Thornton Heath to Crown Hill and Addiscombe and from Croydon High Street to the Red Deer in South Croydon; however, the High Street was too narrow for this section to be linked to the other two routes and significant demolition and road widening would have been needed. Despite this deficiency, the system was profitable and well supported. However, things took a turn for the worse when the Norwood & District Tramways Company (N&DTCo) arrived on the scene. The N&DTCo planned to build a small tramway network around South Norwood, which would connect with the lines of the CTCo, but could not raise enough capital for construction to start. This led the N&DTCo to approach the CTCo with a view to merge the two companies and then expand their networks as planned. The CTCo agreed to such a merger, a decision that it would later live to regret, and a new company known as the Croydon & Norwood Tramways Company (C&NTCo) came into being on 2 August 1883.

HORSE TO STEAM AND GAS

Following the merger, the C&NTCo leased operation of the network to the Steam Tramways Traction Company (STTCo), a subsidiary of the City of London Contract Corporation that would prove to be the undoing of many UK tramway companies. The STTCo stated that it aimed to introduce steam-operated services, at a time when many other tram operators were also experimenting with this method of propulsion as a potential replacement for horse trams. However, its efforts would be a complete failure with steam tram engines only being used for trials rather than in revenue-earning service. The STTCo finally pulled out on 31 December 1884, with the C&NTCo taking over operations itself from the following day. However, the company was in difficulties having spent all its capital on building lines that proved to be unprofitable. The C&NTCo tried to stay afloat for as long as it could, surviving by simply not running services on some lines, but a liquidator was finally appointed on 25 October 1887.

A new tramway company was eventually formed, which was effectively the old CTCo minus its former partner. The new company also traded under the name Croydon Tramways Company even though it was a de facto newly established operator. In its new incarnation CTCo took over the assets of the former C&NTCo from the liquidator on 1 January 1890, running services on what were broadly the former routes of the original CTCo. The two halves of the system were finally connected on 6 January 1897 although the new line was built and owned by, and leased from, Croydon Corporation. Meanwhile in 1891 and 1892 two trials with battery electric trams took place, then in the autumn of 1893 Croydon experimented with a gas tram designed by Carl Lührig in Germany. Similar gas tram trials took place around this time in Trafford Park in Manchester and in Blackpool and Neath.

Chapter 1: Croydon's First Trams: 72 Years of Service

Right: Car 51 on a Thornton Heath-bound service in about 1928. This tram was built by Milnes in 1902 and was rebuilt in 1928–30 with a top cover and two trolley poles, one for each direction of travel. It was then renumbered 26 by Croydon Corporation Tramways (CCT), then to 370 by the London Passenger Transport Board (LPTB) in 1933. *W.J. Haynes collection*

Below: South Metropolitan Electric Tramways (SMET) type K car 18 in central Croydon showing Tooting & Croydon on its front destination blind and Croydon–Mitcham–Tooting on its side destination board, probably in about 1925. Car 18 was built by Milnes for CCT in 1902 and was transferred to SMET in 1906. By the time SMET and its assets were transferred to the LPTB in 1933 this tram was in store at Fulwell depot and was scrapped the following year. *Frank Merton Atkins collection*

Lighting Company (SMET) was formed (not to be confused with outer north and west London/Middlesex/Hertfordshire area operator Metropolitan Electric Tramways), which had ambitious plans to build its own tramway network. Services operated by the SMET included routes from West Croydon to Carshalton,

Below: Car 346 outside Penge depot. The destination blind reads Robin Hood, referring to the now demolished Robin Hood pub, and the side destination board reads "5 – Crystal Palace–Anerley Station–Robin Hood–Selby Road–Norwood–Selhurst–West Croydon". Adjacent to the staircase nearest the camera it carries an advert for the 1933 film Little Women starring Katharine Hepburn, showing at the Croydon Empire on Monday 21 May, so this photo dates from 1934. This tram was originally CCT W/1 class car No. 2, built by Brush in 1911. It was fitted with track brakes to enable it to run over Anerley Hill. In 1933 it was renumbered 346 when CCT and its assets became part of the LPTB. *D.W.K. Jones collection*

GAS TO ELECTRIC

However, this lease did not last long, as the Corporation had plans to build its own electric tramway system and, probably with this end in mind, it acquired the CTCo on 22 January 1900. Despite this acquisition, the Corporation chose not to operate the tramway itself but instead leased it to the British Electric Traction Company (BETCo), which owned, part-owned or leased some 50 tramway concerns all over the British Isles. The first passenger-carrying electric trams ran on the Norbury–Purley route on 26 September 1901 with the last horse-drawn services thought to have operated in late January or early February 1902.

In 1906 Croydon Corporation took back control of the tramway from BETCo. Two years earlier, in 1904, the South Metropolitan Electric Tramways &

CROYDON: TRAM TO TRAMLINK

Sutton, Mitcham and Tooting, opened in 1906. The Croydon Corporation route from Croydon to Penge opened in two stages in April and May 1906 and was operated by SMET cars until 24 June 1907. Meanwhile from 1906 onwards Croydon Corporation ran the town's existing tram services until they, along with the SMET routes, were absorbed into the rest of London's tram network by the newly formed London Passenger Transport Board in 1933. In its heyday between the early 1900s and the late 1920s, the Croydon network included routes to Purley, Addiscombe, Crystal Palace, Penge, Sydenham, Thornton Heath and Norbury along with the SMET routes to Tooting, Mitcham, Wallington, Carshalton and Sutton. Also, from 7 February 1926 route 18 ran through from the Victoria Embankment in London's West End to Croydon, thus offering a cheaper if somewhat slower alternative to taking the train to central London.

Above: The more observant may notice this reminder of Croydon's original tram network in the form of a Croydon Corporation Electric Tramways manhole cover in Lower Addiscombe Road, taken on 13 January 2012. *Keith Fender*

Above: Covered top eight-wheel car 22E on Thornton Heath–Croydon route 42 sometime between 1933 and 1936–37 when this class of tram was withdrawn. These trams were built as open toppers by Milnes in 1902 with their top covers being fitted in 1928 and the E suffix added to their numbers by the LPTB in 1933. *M.J. O'Connor/National Tramway Museum collection*

Above: London Transport E/1 class car 393 in about 1950 showing Greyhound – Croydon on the destination blind and Thornton Heath–West Croydon–Greyhound on the side destination board on the lower deck. No. 393 was ex CCT car 49, built in 1928 by Hurst Nelson. *Lens of Sutton collection*

Above: B1 class trolleybus 64 outside the famous Crystal Palace on a trial run for route 654 which would start running here from West Croydon on 9 February 1936. The service would then operate for 23 years until closure in 1959, but this photo would be unrepeatable after 30 November 1936 when the Crystal Palace was tragically destroyed by a fire. *British Trolleybus Society collection*

DECLINE, CLOSURE, BUS & TROLLEYBUS REPLACEMENT

Nonetheless, the late 1920s also saw the beginning of the end of the tramway era, as by this time the petrol engine was becoming more efficient and the comfort and reliability of motor buses was improving. The war effort also meant that maintenance was neglected during World War I, and the Addiscombe line closed and was replaced by buses in 1927, then in the mid-1930s the Mitcham/Tooting and Wallington/Carshalton/Sutton routes were converted to trolleybus at a time when many other tramways across the UK were being superseded by this mode of transport. Trolleybus route 630 ran all the way from West Croydon to Harlesden and was launched on 12 September 1937, closing 23 years later on 19 July 1960, two years before the end of London trolleybuses in 1962. Croydon's other trolleybus route, the 654, was inaugurated in two stages:

Chapter 1: Croydon's First Trams: 72 Years of Service

between Sutton and West Croydon on 8 December 1935 and between West Croydon and Crystal Palace on 9 February 1936, with the entire route closing on 3 March 1959. Meanwhile on 7 April 1951 the last of Croydon's first generation electric trams ran between Purley and Thornton Heath, just over a year before the closure of London's remaining tramway on 5 July 1952. It would then be almost half a century before trams returned to the streets of Croydon.

RAPID TRANSIT TRAMWAY PROPOSAL

However, despite (or perhaps because of) the imminent demise of London's trams, in 1950 a group called the Tramway Development Council drew up a radical proposal for a "rapid transit tramway" running along a new corridor from Purley and Croydon to a new transport interchange at Kennington Oval, featuring a mixture of on-street tracks, dedicated rights of way, tunnels and flyovers. This would have been operated by twin-coach trains known as "rail coaches" and would have connected with the Northern Line. An extension of the rapid transit line to Victoria, Marble Arch and Paddington was also considered. Most stations along the route would have consisted of basic platforms with shelters, and there would have been larger stations with better facilities where the line interchanged with the Underground or suburban heavy rail network including an underground tram station in Brixton where Acre Lane and Coldharbour Lane cross Brixton Road and Brixton Hill. This scheme never saw the light of day, and it would be 50 years before a similar project would materialise, but it could

Above: K2 class trolleybus 1202 departs for Harlesden on route 630 on 2 July 1960; behind it is K2 1327. Alongside is a member of the Country bus department of London Transport – RT 3148 on route 409 to Godstone garage. This is a scene that would soon disappear as trolleybuses were replaced by motor buses on 20 July 1960. *Peter Mitchell*

Above: 2-EPB EMU 5774 is seen near Beddington Lane on a Wimbledon–West Croydon service sometime in the 1950s. *Chris Wilson collection*

be described as an early version of the Tramlink project except that the present day network has no underground running.

THE TWO FORMER HEAVY RAIL ROUTES

For the sake of completeness, a brief history of the two former heavy rail routes that now form part of the Croydon tram network (three if we class the Addiscombe branch as a separate route) should also be included here. Both lines were something of a passenger backwater throughout the entire period of their existence.

Above: 456 024 in Connex yellow and white livery departs from Waddon Marsh with the 13.35 West Croydon–Wimbledon on 31 May 1997, the last day of heavy rail services on the line. Turners Way Gas Works can be seen on the left. *Alex Dasi-Sutton*

CROYDON: TRAM TO TRAMLINK

Left: 456 023 in Network SouthEast livery departs from West Croydon with the 11.36 to Wimbledon on 11 January 1992. Just visible on the left are the Croydon (left) and Crystal Palace (right) radio and TV transmitters, which dominate the skyline for miles around. *Alex Dasi-Sutton*

Wimbledon–West Croydon

The former Wimbledon–West Croydon line has its origins in the Surrey Iron Railway, a horse-drawn plateway that ran between Wandsworth and Croydon via Mitcham. This was opened in two stages in 1802 and 1803 by the Surrey Iron Railway Company to carry minerals, mainly building materials, coal, corn, lime, manure and seeds. It was a public toll railway on which independent hauliers could use their own horses and wagons. However, after a brief spell of financial success, the line lost much of its traffic after the Croydon Canal opened in 1809 and suffered further with the closure of the underground stone quarries at Merstham in the 1820s. These were served by the Croydon, Merstham & Godstone Railway, another mineral railway built as an extension of the Surrey Iron Railway. Although the Surrey Iron Railway continued to cover its costs, it was unable to update its technology or keep the track in good condition and closed in 1846.

Nine years later, in 1855 the Wimbledon & Croydon Railway opened the Wimbledon–West Croydon line over part of the trackbed of the Surrey Iron Railway. Wimbledon had already been linked by rail to central London since the opening of the London & Southampton Railway (which later became part of the London South Western Railway (LSWR)) in 1838, and the following year the London & Croydon Railway opened between London Bridge and West Croydon. There was soon pressure for a railway linking Wimbledon with Croydon and serving the industries in the area, and in 1853 the W&CR obtained an Act of Parliament to build an 11-mile line from Wimbledon to Epsom, linking the LSWR with the London, Brighton & South Coast Railway (LB&SCR). This scheme was changed to a 5.75 mile line from Wimbledon to West Croydon, which opened on 22 October 1855.

The line was initially operated under contract by its engineer George Parker Bidder, a railway engineer who was also involved in the construction of parts of the Great Eastern Railway. After only a year of operation, in 1856 the route was leased to the LB&SCR, which bought the line outright in 1866.

When the line first opened it had intermediate stations at Morden (later known as Morden Halt, then Morden Road from 1951), Mitcham and Beddington Lane. These were joined in 1868 by Merton Park and Mitcham Junction stations, which opened at the same time as the Merton Park–Tooting and Wimbledon–Streatham South Junction lines, which together formed the Tooting, Merton & Wimbledon Railway, and the Peckham Rye–Sutton line (the South London & Sutton Junction Railway), and 62 years later in 1930 Waddon Marsh opened to serve new housing in Waddon, Croydon Gas Works, and Croydon A & B Power Stations. The Tooting, Merton & Wimbledon Railway formed a loop starting and finishing at Tooting (then known as Tooting Junction); however, the Merton Park–Tooting section, also known as the Merton Abbey branch, closed to passengers in 1929 as a result of competition from tram services but remained open for freight until 1975.

Most of the Wimbledon–West Croydon line was single track throughout its existence but there were originally crossing loops between Mitcham and Mitcham Junction and between Waddon Marsh and West Croydon. Passenger services initially consisted of six trains in each direction on Mondays–Saturdays and two on Sundays. Service levels gradually increased over the years, and for a time in the steam era some trains were extended from West Croydon to Crystal Palace. For most of its existence the line was a fully self-contained operation, however.

Steam push-pull working began in 1919, and in 1930 the line was electrified on the 750 V DC third rail system already in use across much of the rest of the Southern Railway suburban network. Freight traffic on the line declined from the 1950s onwards, and in time so did passenger traffic. In the 1950s and early 1960s, the line had a 20-minute service frequency during the day on Mondays–Saturdays and a half-hourly evening and Sunday service. However, by the late 1970s the loss of freight traffic and falling ridership had led to a reduction in service levels to half-hourly all day on Mondays–Saturdays with no evening or Sunday service. The ending of freight traffic also led to the loop at Waddon Marsh being taken out of use, and in 1971 a landslip at Mitcham led to the loop here being cut short at Mitcham Junction.

For the entire period of electric operation services were generally worked by single 2-car EMUs. Between the 1960s and the early 1990s a Class 416 2-EPB EMU was normally used, and from the start of the 1991–92 winter timetable on 30 September 1991 until the end of heavy rail operation on 31 May 1997 these were replaced by 2-car Class 456 EMUs. By the time the 456s arrived just one unit operated a shuttle service on the line rather than two units that passed each other at Mitcham Junction. This meant that the line latterly had a 45-minute interval service.

Elmers End–Addiscombe/Sanderstead

In 1861 the Mid Kent and South Eastern Railways jointly promoted a Parliamentary Bill for a branch line from New Beckenham to Croydon terminating on Lower Addiscombe Road. The Bill received the Royal Assent on 17 July 1862 and the branch opened on 1 April 1864. This formed an extension to the existing Mid Kent Line, the initial section of which had been opened by the Mid Kent & North Kent Junction Railway between Lewisham and Beckenham Junction in 1857. A proposed extension beyond Addiscombe to Redhill was opposed by the LB&SCR and was dropped. In 1871 an intermediate station opened at Woodside (originally known as Woodside & South Norwood) to serve the nearby Croydon Racecourse which, however, closed as early as 1890 after which usage of this station decreased. Then in 1882 the Elmers End–Hayes section of the Mid-Kent line was completed by the

Chapter 1: Croydon's First Trams: 72 Years of Service

Right: Class 416 2-EPB unit 5773 calls at Woodside with an Addiscombe–Elmers End shuttle on 27 March 1981. *Robin Ralston*

West Wickham & Hayes Railway but was sold to the South Eastern Railway when this section opened on 29 May that year.

The Elmers End–Sanderstead line was opened by the Woodside & South Croydon Joint Railway on 10 August 1885 to provide a link between the South Eastern Railway's Mid Kent Line at Woodside and the Croydon & Oxted joint (LB&SCR and South Eastern & Chatham) line, and is perhaps best described

Below: A rare visitor to Addiscombe on 2 August 1980 was Class 411 4-CEP EMU 7101 on the Kentish Belle Charity Railtour, organised by BR as a staff training trip taking in a number of London suburban lines on the South Eastern Division. The tour also took in the Woodside–Sanderstead line. *Chris Wilson collection*

as having survived on a knife-edge throughout its 98-year history. There were initially just two intermediate stations between Woodside and Sanderstead at Coombe Lane and Selsdon Road (which also had platforms on the adjacent Oxted line until their closure in 1959). Closure of the line was first proposed as early as 1895, and in the early 1900s Kitson steam railmotors were introduced on the route in an attempt to increase ridership in the wake of growing competition from trams and buses. At the same time two new intermediate halts were opened: at Bingham Road between Woodside and Coombe Lane, and Spencer Road between Coombe Lane and Selsdon Road. These were very basic and were built of old wooden sleepers with no buildings.

Passenger numbers had marginally increased by the 1910s, but this was not enough to offset the operating losses of over £2000 per year. In 1915 the railmotor service was withdrawn as a wartime economy measure and the two new halts were closed. From this date, services were suspended for the most part although diverted trains and specials could still use the line. This was followed by full closure of the line on 31 December 1916.

Following the 1923 Grouping, in which all railway companies operating on what would later become the Southern Region were absorbed into the Southern Railway, the SR relaid the track on the line in 1927 and the line was reopened, not to regular traffic but for use by occasional excursion trains and special services. Then in 1935 the SR once again relaid the track, and this time electrified and fully reopened the line with regular passenger services resuming on 30 September that year.

Left: 2-EPB 5744 departs from Bingham Road with a Sanderstead–Elmers End shuttle on 27 March 1981, just over two years before closure of the line. Today the tramway runs along the same alignment but at street level at this location, whereas the original heavy rail line ran on an embankment as in this view. *Robin Ralston*

CROYDON: TRAM TO TRAMLINK

Left: The entrance to Bingham Road station at street level on 12 May 1983, the penultimate day of service. Here the present day tramway crosses Bingham Road at ground level although the adjacent electricity substation on the right, and part of the station entrance retaining wall next to it, remain in situ to this day. *Brian Garvin*

Below: Unit 5744 is seen again, also on 27 March 1981, this time at Coombe Road on a Sanderstead–Elmers End service. The Sandilands Tunnels, now used by the New Addington branch of the tramway, can be seen in the background. *Robin Ralston*

At the same time Bingham Road Halt was rebuilt as a proper station and Coombe Lane station was also rebuilt and renamed Coombe Road while Selsdon Road became plain Selsdon (despite being located two miles away from its namesake village). Spencer Road halt was never reopened, however.

Unfortunately, electrification failed to boost passenger numbers, and low levels of usage led to a reduction in service levels during and after World War II with off-peak services reduced to a shuttle between Elmers End and Sanderstead from 1949. After this date, only peak hour trains continued to run through to and from London Charing Cross or Cannon Street, then in 1967 the Saturday service was withdrawn and in 1976 the line lost all through trains to and from central London leaving only a peak hour shuttle service, most commuters from the Croydon area preferring

Right: Class 415 4-EPB 5209 and 2-EPB 5720 depart Selsdon with the last ever Sanderstead–Elmers End train on the evening of 13 May 1983. Note the exploding detonators beneath the train, which have traditionally been used to mark the closure of a line. This also brought the rare sight of a Southern Railway design EPB unit (with the distinctive small sausage shaped windows above each door droplight) to the line, with trains having more usually been formed of a single BR design 2-EPB as there were no SR design 2-EPBs allocated to the South Eastern Division at this time. The Shell oil siding can be seen on the left. *Alex Dasi-Sutton*

Chapter 1: Croydon's First Trams: 72 Years of Service

Right: The Shell oil siding at Selsdon remained in use until 1993, meaning that the southernmost stub of the Woodside–Sanderstead line at Selsdon was still used by freight traffic until that date. Here 47238 shunts the oil train from Ripple Lane at Selsdon on 12 September 1987. Note the new housing development under construction, part of which would occupy the site of the former northbound platform at Selsdon station. *Alex Dasi-Sutton*

to use the faster routes from South or East Croydon to London Bridge or Victoria.

Meanwhile, in 1963 the line was proposed for closure in the Beeching Report but this sparked an outcry from the newly formed Croydon Transport Users' Association, which claimed that hardship would be caused to the 650 daily users of the line and that no suitable alternative transport was available (despite the fact that Bingham Road was within walking distance of Addiscombe and East Croydon stations while Coombe Road and Selsdon were only a short walk from South Croydon station). Following several public meetings, the Ministry of Transport rejected the closure plans and the line was reprieved.

However, by the early 1980s usage of the line was estimated at less than 200 passengers per day, the track was in poor condition and resignalling was planned for the Selsdon/Sanderstead area. Low ridership levels meant that British Rail could not justify wholesale renewal of the line's infrastructure. This meant that closure was now inevitable, and the last trains ran on 13 May 1983. For most of the post-war period the Elmers End–Sanderstead shuttle trains were generally formed of a single 2-EPB EMU, but some of the trains were worked by pairs of 2-EPBs in the final week of service with the last train being formed of a 6-car EPB (a 4-EPB coupled to a 2-EPB) to cope with the large numbers of passengers paying their last respects.

Tracklifting began shortly afterwards, and by late 1984 Bingham Road and Coombe Road stations had been demolished. The Shell oil siding at Selsdon remained in use until 1993 so freight trains continued to use the southernmost stub of the line almost as far as the bridge over Croham Road to access this facility until that date, after which the junction with the Oxted line was removed. Despite this, to this day the southbound track and platform and part of the northbound platform at Selsdon remain in situ, albeit rather overgrown. One of Selsdon station's main claims to fame is that it is said to have been the last BR station in Greater London to be lit entirely by gas, with the gas lamps remaining operational until closure. Much of the trackbed between Selsdon and the site of Coombe Road station also remains intact but has largely been lost to nature, apart from the actual Coombe Road station site which has been reused for new housing with the present-day tram route running immediately to the east of the station site.

Above: 466 008 awaits departure from Addiscombe with the 10.13 shuttle to Elmers End on 10 May 1997, three weeks before the closure of the branch. *Alex Dasi-Sutton*

Above: Another view of Addiscombe station, looking south towards the buffer stops, also on 10 May 1997. By this time only Platform 2, visible on the left, was still in use. *Alex Dasi-Sutton*

CROYDON: TRAM TO TRAMLINK

Above: Part of the trackbed of the former Addiscombe branch between Woodside and Addiscombe has been turned into a new public park named Addiscombe Railway Park. *Robert Pritchard*

For the next 14 years after the closure of the Woodside–Sanderstead line the Addiscombe branch continued to operate as a shuttle service from Elmers End, its last through trains to central London having been withdrawn by the early 1980s. Unlike the Sanderstead line, the Addiscombe branch retained an all-day service on Mondays–Saturdays but with no Sunday service, and off-peak trains often ran with very few or no passengers. Services were worked by single 2-EPB units until the last EPBs were withdrawn in 1995, after which 2-car Class 466 Networker EMUs were used.

It is likely that the EMU depot at Addiscombe, opened in 1926 to coincide with the line's electrification, helped to keep the branch open, although the depot closed in 1993 by which time plans for the Tramlink project were already well advanced. In 1996, the line's last full year of operation, the signal box was burnt down, destroying the line's Absolute Block Signalling, after which the branch was reduced to single track operation with only Platform 2 at Addiscombe remaining in use. As with the Wimbledon–West Croydon line the last trains ran on 31 May 1997.

After closure, the South Eastern & Chatham Railway Society drew up plans to preserve Addiscombe station and depot which would become a working railway museum. However, Railtrack and Croydon Borough Council did not offer any support for the project, and a five-year campaign to acquire the site was ultimately unsuccessful. Bellway Homes bought the station site in 2000 with outline planning permission for the East India Way Housing Development with 65 homes which now occupies the site of the former Addiscombe station, demolition of the station having started in late 2001 with the retaining wall being all that now remains of the station. This new housing estate gets its name from the East India Company Military Seminary, which was located nearby. The trackbed beyond has been turned into a new public park, appropriately named Addiscombe Railway Park.

THE NEVER REALISED SOUTHERN HEIGHTS LIGHT RAILWAY

Mention should also be made of the ill-fated Southern Heights Light Railway, a scheme for a line from Sanderstead to Orpington which, if built, would have been served by trains running to and from the Elmers End–Sanderstead line. This project was originally proposed in 1898 in connection with the anticipated suburban development of an area either side of the Kent-Surrey border, which led to the formation of the Orpington, Cudham and Tatsfield Light Railway Company with light railway civil engineer Colonel Holman Fred Stephens as its engineer. The line would have run south from Orpington to a station at Green Street Green before turning south-westwards towards Cudham, then due west to Biggin Hill and southwards to its terminus at Tatsfield.

Although the company became defunct after 1906, the Southern Railway proposed a version of the scheme in the mid-1920s involving a line from Orpington to Sanderstead, again with Colonel Stephens as its engineer, which would have been worked by EMUs on passenger services and steam-hauled goods trains. It was envisaged that the line would be served by a Charing Cross–Charing Cross roundabout service via Elmers End, Sanderstead and Orpington. A Light Railway Order was provisionally granted in December 1928 but lapsed two years later in December 1930 because the necessary capital could not be raised. A new LRO was then applied for in January 1931 and this time a different route was proposed in an attempt to reduce construction costs. However, in July 1931 the SR board decided not to support the new LRO application, then in October of that year Colonel Stephens died meaning that the scheme lost its main promoter.

Nonetheless, by the time the project was abandoned, electrification of the Elmers End–Sanderstead line had already been authorised and thus went ahead as planned, being completed in 1935 as mentioned above despite not serving its originally intended purpose. One can only speculate as to whether either or both of the Elmers End–Sanderstead and Sanderstead–Orpington lines would have survived to this day had the Southern Heights scheme gone ahead, and if so whether Tramlink would then still have been built. It is likely that the long-term viability of both lines would have largely depended on how the surrounding area had developed, particularly just before and after World War II and given that much of the area through which the Southern Heights line would have run is now green belt land and subject to stringent planning restrictions. There is also the question of whether New Addington would have developed to the same extent as it did before and after World War II.

PRESERVED LONDON TRAMS AND TROLLEYBUSES

Although there are no preserved first generation trams or trolleybuses specifically associated with Croydon, a number of London trams and trolleybuses have survived into preservation. Most of the trams that are in working order are to be found at the National Tramway Museum (a.k.a. Crich Tramway Village) at Crich, Derbyshire (plus one, E/1 car 1858, at the East Anglia Transport Museum at Carlton Colville, Lowestoft). A number of vehicles are also on display as static exhibits at the London Transport Museum both at its main site in Covent Garden and at the LT Museum Depot in Acton where three open weekends are normally held each year.

There are currently at least five London trolleybuses in working order: 1348 and 1812 reside at the Trolleybus Museum at Sandtoft near Doncaster, whilst 260, 796 and 1201 form part of the collection at East Anglia Transport Museum where the body of trolleybus 1521 also resides, minus its chassis. 1521 was the last trolleybus to operate under its own power in London on the last day of service on 8 May 1962.

K2 class trolleybus 1253 has been on display as a static exhibit at the London Transport Museum in Covent Garden since the museum first opened here in 1980, while the LT Museum Depot at Acton has Q1 class trolleybus 1768 and A1 class trolleybus 1, which was London's first trolleybus when this mode of transport was first introduced by London United Tramways in 1931.

The table below lists all known examples of preserved first generation London trams in the UK (at the time of going to press the restoration of LCC car 1 at Crich was at an advanced stage).

Mention should also be made of ex-Tramlink works car 058 "John Gardner" based at Crich, which consists of a tractor unit equipped with a hydraulic Atlas crane. The vehicle itself is powered by a four-stroke diesel engine and operates with flat bed trailer 061. Both vehicles were originally built in 1978 by Sollinger Hütte in Germany as track construction and railway maintenance vehicles for the then West German national rail operator Deutsche Bundesbahn. Their original fleet numbers were DB Netz 53 0692 and 0692-3 respectively. They operated in Germany until 2005 and with Tramlink from 2006 until 2009, after which they moved to Crich where they have operated since 2017. Car 058 was named in 2015 after the late John Gardner, a former Tramway Museum Society member who was heavily involved in the early years of the museum and who was known for his passion for rail-mounted cranes.

USEFUL WEBSITES

- British Trolleybus Society: www.britishtrolley.org.uk
- British Trolleybuses database: www.trolleybus.co.uk
- Crich Tramway Village: www.tramway.co.uk
- East Anglia Transport Museum www.eatransportmuseum.co.uk
- London Transport Museum: www.ltmuseum.co.uk

Company and fleet number	Type	Built by	Status
Crich Tramway Village			
LCC 1	Bogie double decker	LCC, 1932	Under restoration
LCC 106	Four-wheel open topper	Dick Kerr, 1903	Operational
LUT 159	Bogie open topper	G.F.Milnes, 1902	Operational
MET 331	Bogie double decker	Union Construction Co, 1930	Operational
LPTB 1622	Bogie double decker	Brush & Co, 1911	Operational
Crich Tramway Village (off-site storage facility)			
LTCo (number unknown)	Four-wheel open top horse tram	LTC, 1895	Unrestored
NMTCo (number unknown)	Four-wheel open top horse tram	NMTCo, 1885	Unrestored
East Anglia Transport Museum, Lowestoft			
LT 1858	Bogie double decker	English Electric, 1930	Operational
London Transport Museum, Covent Garden			
WHCT 102	Four-wheel open balcony	United Electric Car Co, 1910	Static exhibit
LTCo 284	Four-wheel open top horse tram	John Stephenson & Co, 1882	Static exhibit
London Transport Museum Depot, Acton			
MET 355	Bogie double decker	Union Construction Co, 1931	Static exhibit
LCC 1025	Bogie double decker	LCC/Westinghouse, 1907	Static exhibit

CROYDON: TRAM TO TRAMLINK

Above: London Transport E/1 class tram 1858 at East Anglia Transport Museum on 10 September 2006. *Alan Yearsley*

Below: Passengers board LCC four-wheeled open top car 106 at the Town End terminus at Crich Tramway Village on 31 July 2016. Berlin car 223 006-3 is on the right; this is the museum's "access tram" and is equipped with a wheelchair lift. *Alan Yearsley*

Chapter 1: Croydon's First Trams: 72 Years of Service

Above: Metropolitan Electric Tramways (not to be confused with SMET) centre entrance car 331 departs Wakebridge towards Crich Town End on 29 August 2021. These centre entrance cars were sometimes referred to as "Felthams" because they were built by the Union Construction Company based in Feltham, and were a regular sight in the Croydon area. 331 was built in 1930 and moved to Sunderland in 1937 where it became car 100 but has now been restored to MET condition. *Alan Yearsley*

Above: LCC car 1 was undergoing restoration when it was seen on the depot fan at Crich Tramway Village on 21 September 2019. *Alan Yearsley*

CROYDON: TRAM TO TRAMLINK

Above: K2 class trolleybus 1201 at East Anglia Transport Museum on 10 September 2006. *Alan Yearsley*

Above: Ex-Tramlink works car 058 "John Gardner" is seen near the Town End terminus at Crich on 21 September 2019 sandwiched between two Blackpool trams: open "boat" car 236 and car 167 just visible on the right. Leeds car 399 can be seen in the distance at the terminus. *Alan Yearsley*

CHAPTER 2:
THE TRAMLINK PROJECT

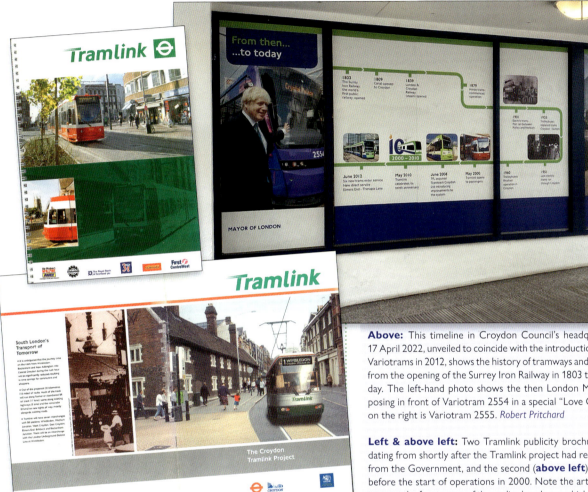

Above: This timeline in Croydon Council's headquarters as seen on 17 April 2022, unveiled to coincide with the introduction of the first Stadler Variotrams in 2012, shows the history of tramways and railways in Croydon from the opening of the Surrey Iron Railway in 1803 to the (then) present day. The left-hand photo shows the then London Mayor Boris Johnson posing in front of Variotram 2554 in a special "Love Croydon" livery, and on the right is Variotram 2555. *Robert Pritchard*

Left & above left: Two Tramlink publicity brochures; the first (**left**) dating from shortly after the Tramlink project had received the go-ahead from the Government, and the second (**above left**) dating from shortly before the start of operations in 2000. Note the artist's impression of a tram on the front cover of the earlier brochure, which was rather different from the design of tram that eventually emerged!

Tramlink grew out of a transport study entitled "Light Rail for London" carried out jointly by British Rail and London Regional Transport in 1986 covering all of Greater London, and a need for better rail services in Croydon and the surrounding area including linking previously separate routes. Whilst most of the network serves areas that already had rail services, there is one important area, New Addington, that had never previously had any kind of rail service.

New Addington was developed as a new suburb in the 1930s and was promoted as a pleasant area free from the stress, noise and grime of inner city areas, and further new housing developments after World War II led to a growth in the area's population to around 25 000 people. However, by the late 20th century these advantages were increasingly seen as pointless without some form of rail-based transport particularly with increasing private car traffic. The area suffered from poor connectivity, with buses to central Croydon taking up to 45 minutes in peak hours for a journey of just 5 miles. Between the 1960s and the early 1980s a number of solutions were considered to provide New Addington with an improved transport service, including a people mover or computer-controlled mini-tram, a monorail and a hovertrain. None of these schemes ever saw the light of day.

A year later, in 1987, BR, LRT and officers from Croydon Council carried out a further study that found that an initial network serving Wimbledon, Elmers End and New Addington would be feasible and worthy of further investigation. This was the year that saw the opening of the first phase of London's other light rail network, the Docklands Light Railway, and with it came the growth of London's Docklands as a major business district while areas such as Croydon, which had seen significant new office development in the 1950s as part of the Government's dispersal of offices programme, were in danger of falling into decline without serious measures to improve their transport links. Croydon Council also realised that a purely roads-based approach would not work, as measures such as road widening would be unpopular with local residents who would have to suffer from increased noise and pollution as well as potentially losing part of their gardens or even having to move house if their properties needed to be demolished.

Also in 1987, Chris Green, the then Managing Director of BR's Network SouthEast sector, published plans to convert the Purley to Caterham and Tattenham Corner branches to light rail and establish a through service from Croydon to Lewisham via Addiscombe and

CROYDON: TRAM TO TRAMLINK

Above: In 1987 BR Network SouthEast Managing Director Chris Green proposed to convert the Caterham and Tattenham Corner branches to light rail. On 4 August 2018 377613 stands at the terminus at Caterham awaiting departure for London Bridge. If this scheme to turn the branch into a tramway had come to fruition then scenes such as this would not be possible, and it is questionable whether even the station building would have survived. *Robert Pritchard*

Hayes. However, even at this time the idea was not new: a private study in 1962 assisted by BR engineers showed that the Wimbledon–West Croydon line could easily be converted to light rail, and during the 1970s several BR managers were aware of the advantages of such a conversion. Around this time Tim Runnacles, the then Director of Light Rail at London Transport, had suggested an electric tramway as a potential solution to the New Addington problem, as recalled by his successor Scott McIntosh who was involved in the early development of the Tramlink project as well as that of the DLR.

In 1988, the London Assessment Studies looked at transport issues in a number of key corridors across Greater London including the A23 at Croydon. Road improvements (never implemented because of public opposition) and public transport-based solutions were recommended, including a light rail system albeit serving a somewhat larger area than what would eventually emerge with the Tramlink scheme.

Croydon Borough Council carried out a survey in 1990 to gauge support for a tramway, with over 50 local businesses and organisations being asked for their opinion. Reaction was generally positive, and the following year the council deposited a Bill in Parliament, which received the Royal Assent in 1994.

BUS OR LIGHT RAIL?

Meanwhile in February 1991 Croydon Council published a report highlighting the need for a radical solution to Croydon's transport problems, without which it warned road congestion and reliability of bus services would get worse. The report did examine a number of alternatives to light rail, including improvements to conventional bus services, guided buses, and more advanced options such as Automated Guided Vehicles (AGVs). However, at this time only two guided bus systems were in operation in the entire world although the technology has since been adopted in a few places across the UK such as Crawley, Ipswich and Leeds. The report also found that guided buses had several disadvantages compared to light rail, including their inability to couple together in peak hours and their need to revert to on-street running in town centres while guided busways can be visually intrusive in parkland and rural areas. It should also be remembered that any form of bus-based system does not have the same ability to attract motorists out of their cars as light rail or other rail-based systems do.

CONSTRUCTION STARTS

Construction work started in 1997 following the closure of the Wimbledon–West Croydon and Elmers End–Addiscombe lines for conversion to light rail. On 31 May that year, their last day of heavy rail operation, train operator Connex ran a "Last In Last Out" railtour which took in both these lines along with the Folkestone Harbour and Newhaven Marine branches. The train, formed of 4-VEP EMUs 3543 and 3544, left Addiscombe and Woodside stations to the sound of detonators to mark the end of an era on the branch. From 2 June a new limited stop bus service numbered TL1 started running between Wimbledon and West Croydon, stopping close to each of the existing stations and proposed new tram stops. This service continued to run until the opening of the Tramlink route to Wimbledon. No such provision was made for Addiscombe branch passengers, as it was presumably considered that the area was already adequately served by existing bus routes.

Close to the junction of the former Addiscombe and Sanderstead lines, a developer had built flats on the former railway alignment on Teevan Close. These had only gained consent on appeal after opposition from both Croydon Council and London Regional Transport (which would be superseded by Transport for London in 2000). However, the development was then compulsorily purchased and demolished, which added an unnecessary £2–3 million to the cost of the Tramlink project. One new-build house also had to be demolished on the site of the old Coombe Road station to make way for the tram line to New Addington.

Above: Tracklaying in progress near West Croydon station, with no less than three 1980s/90s era London buses arriving at the adjacent bus station. The white building in the background is now painted in the London Overground orange and grey colour scheme and houses a secondary entrance to West Croydon station. *John Law*

Left: The closure notice displayed at stations in the area announcing the withdrawal of the Wimbledon–West Croydon heavy rail service. *Geoffrey Skelsey*

Chapter 2: The Tramlink Project

Above: Construction work in progress outside East Croydon station. Note the Network SouthEast logo still visible on the station roof even though NSE officially ceased to exist in 1994 (and Network SouthCentral, which ran the Central Division of the Southern Region as a shadow franchise from that date and also used the NSE logo, was superseded by Connex in 1996). *John Law*

Above: Another view of construction work outside East Croydon station, taken on 18 March 1999 looking towards Croydon town centre. *Keith Fender*

Above: A close-up view of East Croydon tram stop under construction, also on 18 March 1999. *Keith Fender*

CROYDON: TRAM TO TRAMLINK

Above: Tracklaying in progress on George Street on 18 March 1999 at the junction with Wellesley Road, where the Croydon town centre loop from West Croydon can be seen curving off to the right in the foreground. *Keith Fender*

Above & below right: Two views of construction work on Addiscombe Road, both taken on 18 March 1999. **Above:** looking towards East Croydon station; **Below:** looking towards Sandilands. *Keith Fender (2)*

After a few false starts Tramlink finally opened in three stages in May 2000, starting with Route 3 (Croydon–New Addington including the Croydon town centre loop) on 10 May. A party of invited guests were taken to the New Addington terminus on two special buses that morning, and after an opening ceremony by Mayor of Croydon Dr Shafi Khan they then travelled to Croydon on trams 2530 and 2543, which ran one behind the other and left New Addington at about 12.30. Car 2549 formed a third special carrying invited

TRAMLINK OPENS

By the autumn of 1999 work had been completed, and opening was originally scheduled for November of that year but was delayed by a number of technical issues, including problems with the signalling and traffic signals in Croydon town centre causing delays. There were also reports of interference by stray currents from signalling on nearby heavy rail lines meaning that some of the tracks in central Croydon had to be taken up and relaid.

Chapter 2: The Tramlink Project

Right: Two trams outside East Croydon station on 11 March 2000, during a period of "ghost running" before the start of public service. *Geoffrey Skelsey*

enthusiasts from Addiscombe to New Addington via Croydon. Among the party of invited guests were several surviving members of staff from Croydon's original tramways.

After arrival at New Addington, car 2543 worked the first public service to East Croydon while 2550 performed the first public run from East Croydon to New Addington. For the rest of that day passengers were allowed to travel for free in a move to encourage the local population and visitors from further afield to sample the new system. This was followed by Route 2 (Croydon–Beckenham Junction) on 23 May and lastly Route 1 (Wimbledon–Elmers End) on 30 May, again in both cases involving an opening ceremony around mid-morning followed by the start of public services from around 12.00.

EFFECTS ON BUS SERVICES

Bus services in mainland Britain outside London were deregulated in 1986, and on many other tram and light rail systems this has led to wasteful competition between bus and tram/light rail services. However, to this day this policy has never been implemented in London, so the opening of Tramlink led to a reorganisation of bus routes in the areas served so that bus services complemented the trams rather than competing with them. In particular the X30 Croydon–New Addington limited stop service was withdrawn on 10 May 2000, the day that the tram route to New Addington opened, and a month later on 10 June

Above: CR4000 car 2532 waits at New Addington to form a service to central Croydon on 18 May 2000, eight days after the start of passenger services. *Geoffrey Skelsey*

the long-established Woolwich–Croydon route 54 was truncated at Elmers End.

TRAMLINK FEEDER BUS ROUTES

Also to coincide with the opening of Tramlink three feeder bus routes were introduced. Numbered T31, T32 and T33, these served a number of

Left: Another view of New Addington terminus also on 18 May 2000, with car 2548 waiting to depart for Croydon. *Geoffrey Skelsey*

CROYDON: TRAM TO TRAMLINK

neighbourhoods of New Addington and the surrounding area that were not close to the tram route. All three routes served a new bus/tram interchange at Addington Village tram stop. Routes T31 and T32 were initially operated by First Orpington Buses/Centrewest Orpington and worked by buses that carried the same red and white livery as the trams, while route T33 was operated by Metrobus for most of its life as the T33. This is the only Tramlink feeder route still in existence today, albeit renumbered as the 433 in 2015 when the T31 and T32 were withdrawn and partly replaced by changes to other bus routes.

When first introduced, bus routes T31, T32 and T33 offered through ticketing with Tramlink meaning that passengers had only to buy one ticket for their entire journey from a bus stop to any tram stop or vice versa. Today, anyone making such a journey is catered for by the Hopper Fare (see fares and ticketing, chapter 6).

Above: Car 2552 at Beckenham Junction on 4 June 2000. Rather oddly this tram is showing Beddington Lane on its destination blind; at this time trams from Beckenham Junction normally started and terminated there and took a complete circuit of the Croydon town centre loop. Also, some Wimbledon-bound trams start from Beddington Lane, but Therapia Lane is normally used as a terminating point for late evening trams to avoid the need to run empty between Wimbledon or central Croydon and the depot. *Geoffrey Skelsey*

MAKING A DIFFERENCE?

In June 2002 TfL and the Department for Transport published the Croydon Tramlink Impact Study, which looked at the effects of Tramlink on travel behaviour compared to before the system opened. Data for all areas served by the network showed that Tramlink accounted for 15% of all journeys on weekdays and 17% at weekends, while the percentage of trips by car, bus and on foot had fallen and only rail (including London Underground) had remained constant.

Although 69% of Tramlink users had previously travelled by bus, 7000 car journeys per day had transferred to the trams. Also, 22% of tram passengers were making journeys that they had not previously made at all, although in many cases this had not been a necessary journey anyway. Of all passengers who had switched to Tramlink, 55% had the use of a car and 86% had access to an alternative mode of transport.

Above: Car 2536 at Wimbledon on 4 June 2000, five days after the Wimbledon line opened. Until the tram terminus at Wimbledon was rebuilt in 2015, all trams simply terminated at one end of Platform 10. *Geoffrey Skelsey*

The Wimbledon–Croydon section had seen an eightfold rise in patronage during the morning peak between 1994 (when it was still a heavy rail line) and 2001, Tramlink's first full year of operation. Higher service frequency and a wider range of journey opportunities had contributed to this increase.

OWNERSHIP CHANGES

In December 1994 Transport Secretary Brian Mawhinney announced public funding for the Tramlink project along with Birmingham's Midland Metro (now West Midlands Metro) on condition that both schemes be progressed as a Private Finance Initiative (PFI) in line with Government policy. Transport Minister Steven Norris then formally launched a competition to develop the Tramlink scheme in May 1995, with an advert being published in the Official Journal of the European Union to start the tendering process. London Regional Transport invited bids for the 99-year concession to design, build, operate and maintain the Tramlink system. A total of eight applications were received, and these were then reduced to a shortlist of four:

Altram:	John Laing, Ansaldo Transport and Serco;
Croydon Connect:	Tarmac, AEG, Transdev;
CT Light Rail Group:	GEC Alsthom, John Mowlem & Co, Welsh Water;
Tramtrack Croydon:	First Centre West Buses, the Royal Bank of Scotland, financial services company 3i, Sir Robert McAlpine, Amey Construction, Bombardier Eurorail.

Chapter 2: The Tramlink Project

In April 1996, Tramtrack Croydon Ltd (TCL) was announced as preferred bidder, and was awarded a 99-year concession in November of that year. A new subsidiary of First Centre West responsible for tram services, known as First Tram Operations Ltd (TOL) was then formed. TCL won the competition because it required the lowest level of Government funding, and because of this the consortium agreed to a number of cost-cutting measures including terminating the New Addington line at the north end of Central Parade rather than continuing to New Addington Library, replacement of a proposed tunnel at the top of Gravel Hill on the New Addington line by a road crossing, and a reduction in the length of the double track section on the Wimbledon line (although this would later be remedied by the doubling of part of this route in 2012). This also meant that cheaper overhead electrification masts were used and a cheaper ticket machine design was initially chosen that would only issue non-encoded tickets although in the event the machines that were eventually selected did issue magnetically encoded tickets as these would be needed to operate the ticket gates at Wimbledon station where the tram platform would be located inside the gates (and holders of One Day Travelcards bought at tram stops would likely also need to pass through ticket gates elsewhere).

However, by early 2003 TCL was in serious financial difficulty after an over-optimistic revenue forecast. It should be noted that at the time that TCL was awarded the concession demand forecasts had suggested that 25 million passengers per year would be carried, but in the event this level was not reached (and exceeded) until 2007–08 (see ridership figures, chapter 6). It had also taken the company about 18 months from the opening of Tramlink to reach the level of service originally envisaged. It was reported that banks and shareholders backing TCL were understood to be considering a debt-for-equity swap in an attempt to refinance Tramlink. In 2004, TCL agreed a refinancing package with its banking group. However, in January 2005 TCL threatened to take legal action against TfL, claiming that TfL's fare structures had made travel on buses cheaper than on the trams during Tramlink's first few years of operation.

London Mayor Ken Livingstone also claimed in early 2007 that TCL had not shown any initiative nor willingness to improve capacity on Tramlink through an increase in services, and deplored the then recent timetable changes which had seen a reduction in service levels on the New Addington line. Because of this, Livingstone asked TfL to provide extra bus services, particularly on route 130 to and from New Addington, to help alleviate overcrowding on some of the busiest sections of Tramlink. At around the same time TCL lost a court case brought against TfL subsidiary London Bus Services seeking compensation on a cash fare basis for new tickets and passes introduced on Tramlink, shortly after Livingstone had called for TCL to lose the Tramlink operating concession.

Eventually, on 27 June 2008 ownership of the Tramlink system was transferred to TfL after TCL shareholders accepted TfL's offer to purchase the business for £98 million. Under the terms of the deal TfL retained all revenue generated by Tramlink and would no longer need to pay compensation to TCL to keep fares low. One month later, from 20 July TfL increased the off-peak service on the Elmers End and Beckenham Junction routes from two to four trams per hour and pledged to run additional trams on the New Addington line to relieve crowding.

Shortly after TfL's takeover of Tramlink from TCL, in October 2008 a new green, blue and white colour scheme was unveiled for the tram fleet to replace the previous red, black and white livery. At the same time Croydon Tramlink was rebranded as London Tramlink (and often referred to simply as Tramlink) to reflect TfL's long-term ambitions (so far never realised) to expand the network beyond the Croydon area. In recent years, Tramlink has officially been rebranded as London Trams.

Above: Shortly after the change of ownership, TfL unveiled a new livery for Tramlink to replace the original red, black and white colour scheme. Car 2552 was the first to be outshopped in the new livery, seen between Sandilands and Lebanon Road on a Wimbledon service on 8 October 2008. *Keith Fender*

CROYDON: TRAM TO TRAMLINK

CHAPTER 3:
LONDON TRAMS
ROUTE BY ROUTE

Above: CR4000 car 2534 calls at East Croydon station with a New Addington service (left) while Variotram 2565 waits to depart with a service to Therapia Lane on 18 March 2022, with the iconic One Croydon office building (formerly known as the LNA Tower) on the right. Normally only late evening services terminate at Therapia Lane, but presumably on this occasion there was disruption on the Wimbledon line or other exceptional circumstances. *Alan Yearsley*

Above: Variotram 2555 is seen shortly after departure from East Croydon on a New Addington service on 11 February 2020. *Jamie Squibbs*

LONDON TRAMS FACT PANEL

Population (Croydon town area):	192 064[1]
System length:	28 km (17 miles)
Number of stops:	39
Number of vehicles:	36 (24 Bombardier CR4000s, 12 Stadler Variotrams)

[1]Figure based on the 2011 census returns. This figure covers the 13 electoral wards that make up the town of Croydon, with the entire London Borough of Croydon having a population of 384 837 (2011 census). The exact size of the population served is difficult to measure, as the system also serves a small area of the Boroughs of Bromley and Merton, with two tram stops (Beddington Lane and Therapia Lane) being located in the Borough of Sutton.

Chapter 3: London Trams Route By Route

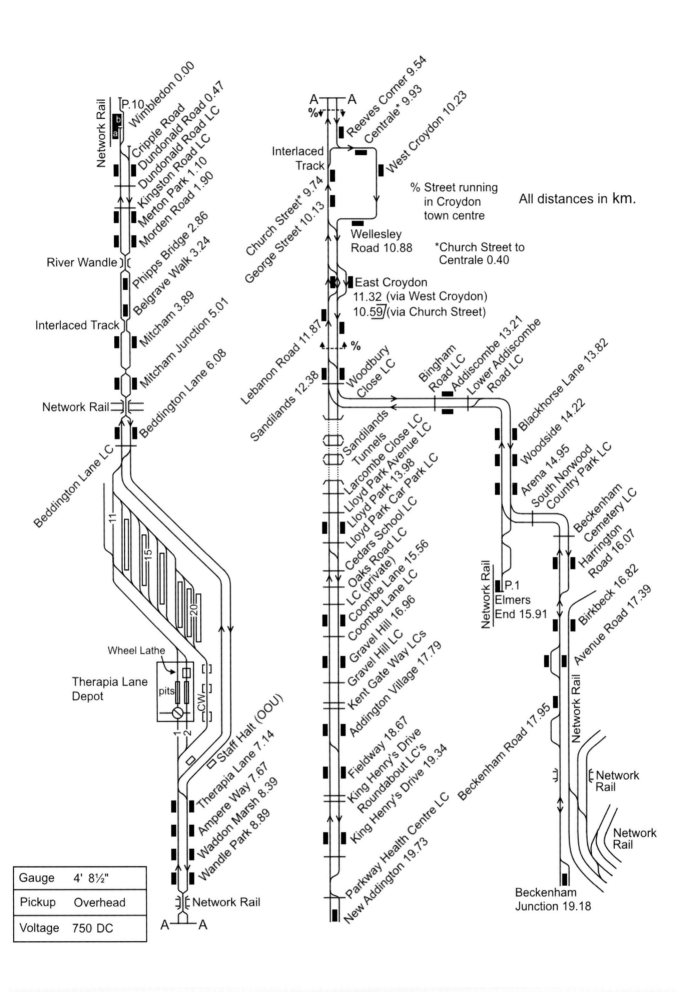

Above: A line map showing the London Trams routes, kilometre distances between stops and the layout of Therapia Lane depot and stabling sidings. *Martyn Brailsford/Branch Line Society*

CROYDON: TRAM TO TRAMLINK

Right: Variotram 2558 approaches Lebanon Road with a Wimbledon service on 18 March 2022. Lebanon Road stop has staggered platforms; the eastbound platform can be seen on the left while the westbound platform is behind the camera. *Alan Yearsley*

Visitors to the area will most likely have their first encounter with the system at East Croydon, West Croydon, Mitcham Junction, Beckenham Junction, Elmers End or Wimbledon station if arriving by train. Of these, East Croydon is the only tram/rail interchange served by all routes, and can be reached from a number of locations all over London and the South-East, so it would seem appropriate to start our tour here. All directions are given assuming that the passenger is facing the direction of travel and travelling away from East Croydon (apart from on the central Croydon loop, which starts and finishes at East Croydon).

NEW ADDINGTON

As the New Addington line has the largest section of completely new-build route, and serves the only destination that has never previously been served by any form of rail-based transport as mentioned in chapter 2, we will explore this route first. Whilst waiting for our tram at East Croydon it is likely that we will see a few trams bound for the other destinations on the network (Wimbledon, Beckenham Junction and Elmers End).

On departure from East Croydon we run along a short section of on-street track on a gentle uphill gradient as far as the first stop, Lebanon Road, the only stop on the network with staggered platforms, the outbound stop being opposite Lebanon Road bus stop. We pass the westbound platform on our right immediately before arriving at the eastbound platform. After Lebanon Road the tramway is in the road until we cross the A232 at a light controlled crossing after which we are on reserved track as far as the next stop, Sandilands.

Between Lebanon Road and Sandilands the A232 Chepstow Road trails in on the right; the A232 then becomes the Addiscombe Road as it heads west. For several years at the junction with Chepstow Road there were problems with cars trying to follow trams onto the tram tracks towards Sandilands, with Croydon Council having been in a long-running dispute with the Department for Transport as to the appropriate signage at this location as motorists appeared

Below: CR4000 car 2530 is seen between Lebanon Road and Sandilands with an Elmers End service on 17 April 2022. *Robert Pritchard*

Below: CR4000 car 2546 arrives at Sandilands with a New Addington service on 17 April 2022. *Robert Pritchard*

Chapter 3: London Trams Route By Route

Above: Variotram 2556 is seen shortly after emerging from Sandilands Tunnel, which can be seen in the background, as it approaches the site of the former Coombe Road station with a New Addington service on 17 April 2022. *Robert Pritchard*

Above: Variotram 2564 passes the site of the former Coombe Road station at Larcombe Close with a New Addington–Wimbledon service on 15 September 2017. The routes were reorganised five months after this photo was taken, since when most New Addington services have run only to and from central Croydon with only two early morning workings continuing to Wimbledon. In the distance can be seen the chevron warning signs that were installed following the 2016 Sandilands derailment to warn drivers of the sharp curve. *Tony Christie*

CROYDON: TRAM TO TRAMLINK

Above: CR4000 car 2538 curves away from the route of the former Elmers End–Sanderstead line towards Lloyd Park, the first stop after Sandilands Tunnel, with a New Addington service on 17 April 2022. *Robert Pritchard*

not to understand the blue "bus only" or "tram only" signs that were used here but until recently the DfT rejected the use of "No Entry" signs. Thankfully this issue has now been resolved, with the previous blue signs having been replaced by "No Entry" signs with an "Except Trams" sign underneath.

There is a crossover at the Croydon side of the Sandilands stop. On leaving Sandilands we take a sharp right turn onto the trackbed of the former Elmers End–Sanderstead heavy rail line, which closed in 1983. Trams heading for Beckenham Junction or Elmers End turn left at this point. This was the site of the terrible accident in November 2016 when westbound tram 2551 left the rails and overturned as it was going too fast for the junction and seven people lost their lives (see chapter 6). We then almost immediately pass through Sandilands Tunnel, although strictly speaking this consists of three tunnels with very short gaps between them. From the north end to the south end these are Radcliffe Road (Woodside), Park Hill and Coombe Road tunnels, which are 243 metres, 112 m and 144 m long respectively. These are the only tunnels on the network, and were refurbished including the installation of lighting when Tramlink was being constructed for ease of evacuation of passengers in an emergency. The longest distance between stops on the network is between Sandilands and Lloyd Park (1.6 km), in fact the three longest gaps between stops are found between the first four stops on the New Addington line.

Emerging from the tunnels into a tree-lined cutting, we then run alongside Larcombe Close, a new residential street on the site of the former Coombe Road station on our right, and then take a sharp left turn away from the former heavy rail alignment onto the present day New Addington route. The more observant passenger may notice the former railway embankment immediately to the right of Campden Road, on the south side of the A212 Coombe Road, the site of the former Coombe Road station having changed out of almost all recognition as a result of new housing developments.

After leaving the former Elmers End–Sanderstead alignment, we soon arrive at the next stop, Lloyd Park, with the park itself and Lloyd Park Lounge Café to the left. This is an excellent location for photographing trams at various times of day. At this point the line begins to have a more rural feel about it, with Lloyd Park and then Coombe Park and Addington Hills heathland and woodlands taking up much of the area to our left while on our right high density housing gives way to lower density residential areas. There is a crossover just after the Lloyd Park stop – running alongside the A212 Coombe Road and a parallel footpath and climbing gently and then more sharply there is then another long gap between here and the next stop, the semi-rural Coombe Lane, as the line heads in a south-easterly direction. From the end of the former heavy rail alignment until just east of Coombe Lane stop, the line runs on the north side of the A212, but shortly after this point the line crosses over Coombe Lane and runs on the south side of this road, which then becomes Gravel Hill after which the next stop is named. Look to the left just before trams cross the A212 at the crest of the hill and you will see one of the more unusual landmarks to be seen from the tram – the Thames Water Addington Water Tower. Located to the right in these leafy suburbs are the Royal Russell School and then the Heathfield House & Gardens.

After running gently downhill, at Gravel Hill Addington Palace Golf Club is located on our left, and after leaving Gravel Hill stop the line once again crosses over the A212 road and then runs alongside Addington Park as far as the end of Gravel Hill – trams can reach some of their highest speeds on this downhill section. There are some great spots for photography around here, the New Addington line certainly being the best for linesiding trams. Gravel Hill is shown as the stop for John Ruskin College, just a few minutes walk away. We next take a sharp left turn onto the Kent Gate Way, passing the entrance to Addington Park, where the line runs between the

Chapter 3: London Trams Route By Route

Right: CR4000 car 2530 is seen just after leaving Lloyd Park with a New Addington–Wimbledon service on 15 September 2017. Car 2545 is just visible in the background at Lloyd Park stop with a New Addington-bound service. *Tony Christie*

Below: CR4000 car 2546 arrives at Lloyd Park with a New Addington–Croydon service in the early evening sunshine on 18 March 2022. *Alan Yearsley*

northbound and southbound carriageways in a north-easterly direction almost as far as the next stop, Addington Village. As its name suggests, this stop serves a much smaller and longer-established settlement than its quasi namesake at the far end of the line. Addington Village stop doubles as a tram-bus interchange with the bus station being located on our left.

On leaving Addington Village there is another crossover and the line once again heads in a south-easterly direction and climbs again alongside Lodge Lane as far as the next stop, Fieldway. About halfway between these two

Below: Car 2540 on the scenic section of the New Addington Line past Addington Hills as it climbs towards Coombe Lane with a New Addington service on 17 April 2022. *Robert Pritchard*

CROYDON: TRAM TO TRAMLINK

Above: Variotram 2556 arrives at the tranquil semi-rural Coombe Lane stop with a New Addington service on 17 April 2022. *Robert Pritchard*

Left: Another view of Coombe Lane stop, looking towards New Addington, shows car 2541 calling with a New Addington–Croydon service, also on 17 April 2022. Note the Thames Water Addington Water Tower just visible to the right of the tram. *Robert Pritchard*

Below: Car 2549 is seen near Gravel Hill with a New Addington service on 18 April 2022. *Robert Pritchard*

Chapter 3: London Trams Route By Route

Above: CR4000 car 2540 pauses at Gravel Hill bound for New Addington on 17 April 2022. *Robert Pritchard*

Above: CR4000 car 2535 arrives at Addington Village with a New Addington service on 15 September 2017. *Tony Christie*

CROYDON: TRAM TO TRAMLINK

© 2023 Platform 5 Publishing Ltd.

LONDON TRAMLINK

Stations: Wimbledon, Dundonald Road, Merton Park, Morden Road, Phipps Bridge, Belgrave Walk, Mitcham, Mitcham Junction, Beddington Lane, Therapia Lane Depot, Therapia Lane, Ampere Way, Waddon Marsh, Wandle Park, Reeves Corner, West Croydon, Centrale, Church Street, George Street, Wellesley Road, East Croydon, Lebanon Road, Sandilands, Addiscombe, Blackhorse Lane, Woodside, Arena, Harrington Road, Birkbeck, Avenue Road, Beckenham Road, Beckenham Junction, Elmers End, Sandilands Tunnels, Lloyd Park, Coombe Lane, Gravel Hill, Addington Village, Fieldway, King Henry's Drive, New Addington

Below: CR4000 car 2530 departs from King Henry's Drive on the last stage of its journey to New Addington on 5 April 2019. *Robert Pritchard*

Chapter 3: London Trams Route By Route

Above: Variotram 2561 awaits departure for central Croydon at the New Addington terminus on 5 April 2019. Note that it is still showing New Addington on its destination display; it could be argued that this is correct, given that most trams on the New Addington line start and finish their journey at New Addington and do a full circuit of the Croydon town centre loop. It is unlikely that many passengers will both board and alight at New Addington and stay on the tram for its entire journey, however! *Robert Pritchard*

stops the nature of the line changes from semi-rural to residential with areas of New Addington starting to appear on our left, and after Fieldway there is one more intermediate stop, King Henry's Drive, before we reach journey's end at New Addington. The track layout on the approach is slightly unusual and there is a very short single track section just before the terminus, which consists of an island platform (trams usually use the right hand platform during normal operations). There are a number of shops close by, including a Boots pharmacy and a fish & chip shop.

BECKENHAM JUNCTION/ ELMERS END

Taking a left turn at Sandilands this time, the line immediately passes under the A232 Addiscombe Road. We are again running along the former Elmers End–Sanderstead heavy rail alignment as far as the third stop, Woodside. In the meantime there are two further intermediate stops – the busy Addiscombe and Blackhorse Lane. After Addiscombe stop, trams cross the A222 Lower Addiscombe Road on a level crossing and there is then a crossover. Also worth noting is the "Tram Stop" café, which gets its name from the former Croydon Corporation tram terminus nearby. At Blackhorse Lane a long ramped access path gives access to the stop. You can get off here and walk a short way along Blackhorse Lane itself (where the bridge over the old railway had to be raised for trams) to see the former Addiscombe railway line, now converted to a footpath and cycleway. Just before Woodside it is possible to work out where the heavy rail branch to Addiscombe diverged on our left. However, a new housing estate and public park now occupies the former trackbed of the Woodside–

Above: CR4000 car 2535 negotiates the curve onto the Beckenham Junction and Elmers End line at Sandilands with an Elmers End service on 25 January 2021. The chevron warning sign can be seen on the left. *Keith Fender*

CROYDON: TRAM TO TRAMLINK

Addiscombe branch after its closure in 1997 as mentioned in chapter 1, and no trace now remains of the old Addiscombe station apart from the retaining wall.

Addiscombe tram stop is located on the opposite side of Bingham Road to the old Bingham Road station. The present day tramway runs at ground level at this location, whereas this stretch of the original heavy rail line ran on an embankment with overbridges over Bingham Road (adjacent to its namesake station) and the A222 Lower Addiscombe Road. With these bridges having been removed, the decision to demolish the embankment and run the tramway at street level eliminated the need to build replacement overbridges or provide lifts or ramps to the platforms at the stops on this section. As the line crosses over Bingham Road, at least in winter when there are no leaves on the trees the more observant traveller may be able to make out a small section of brick wall that formed one side of the stairway to the Elmers End-bound platform of the old Bingham Road station on our left and the electricity substation next to this wall.

Just after leaving Woodside, the street level building of the original Woodside station can be seen on the road overbridge above. This building formerly housed the ticket office but is now derelict. The present day tram stop platforms are located just south of the original station platforms and are accessed via a footpath leading from the former station building. Tram pantographs drop very low as we pass underneath the old railway station entrance.

After Woodside we soon arrive at Arena, which gets its name from Croydon Sports Arena complex, parts of which can be seen at both sides of the tram stop. From here the Elmers End branch continues straight on along the former heavy rail alignment, with South Norwood Country Park on our left (initially behind a high fence) as far as the Elmers End terminus where trams arrive in Platform 1, a terminal bay on the opposite side of an island platform to London-bound Mid-Kent Line trains running from Hayes to Charing Cross or Cannon Street. There is a small station car park to the left of the tram terminus. Before conversion to light rail, this bay platform was used by shuttle trains to Addiscombe (and Sanderstead trains until 1983). Initially double track, the Elmers End branch soon drops to single track, but with a crossing loop just before the terminus. A second tram platform at Elmers

Above: Variotram 2556 passes beneath the street level building of the original Woodside station with an Elmers End–Wimbledon service on 15 September 2017. The original station platforms were located immediately in front of this building, but the tram stop is just behind the camera. *Tony Christie*

Above: CR4000 car 2552 departs from Arena as it heads for Beckenham Junction on 17 April 2022. The Elmers End branch can be seen in the foreground. *Robert Pritchard*

Right: On 8 March 2014 Variotram 2555 awaits departure from the tram bay platform at Elmers End that was formerly used by Addiscombe branch trains (and Sanderstead trains until 1983). On the right Class 376 EMU 376 030 arrives with a Hayes service. Note that at least at the time of this photo the tram platform had tactile paving (as have all tram stop platforms since Tramlink opened) but the heavy rail platforms had yet to be so-equipped. *Tony Christie*

Chapter 3: London Trams Route By Route

Above: Car 2535 is seen again, this time departing from Addiscombe as it crosses the A222 Lower Addiscombe Road also on an Elmers End service, on 17 April 2022. Just behind the rear end of the tram is the Tram Stop café to the right of the William Hill betting shop. The original Elmers End–Sanderstead heavy rail line ran on an embankment at this location, with overbridges over Bingham Road and the Lower Addiscombe Road. *Robert Pritchard*

End to cater for a higher service frequency has been mooted, and probably stands a better chance of seeing the light of day than any other schemes to expand the network (see chapter 7).

Bearing left from Arena, South Norwood Country Park is now on our right as we take a winding curve around the edge of the park. There are some good photographic locations around here on the Beckenham Junction line, with a number of foot crossings, and this is a popular place for walkers and runners. The line takes a sharp left curve before arriving at the next stop, Harrington Road. The large 44-acre Beckenham Crematorium and Cemetery is on our right and continues almost as far as the following stop, Birkbeck; however, there is an empty plot of land that used to form part of the cemetery and could surely be used for new housing.

As the line curves to the right the Crystal Palace–Beckenham Junction heavy rail line joins us, and we then run alongside this line for the rest of the journey. This is an unusual situation where an electrified third rail line parallels an electrified overhead line. Tree-lined, we arrive at Birkbeck station, which has one platform served by the single track heavy rail line and one on the right (as you look towards Beckenham Junction) served by the single track tram line. The next two stops, Avenue Road and Beckenham Road, are both served only by trams. Most of the route from just north of Harrington Road onwards is single track but there is a crossing loop at Avenue Road (used in the normal service) and just east of Beckenham Road. Westbound trams are sometimes held in the loop at Beckenham Road to let an eastbound tram pass.

Above: CR4000 car 2532 passes under Blackhorse Lane as it arrives at the stop of the same name with an Elmers End service on 17 April 2022. This overbridge had to be raised to accommodate the tramway overhead power lines when Tramlink was being built. *Robert Pritchard*

CROYDON: TRAM TO TRAMLINK

Above: CR4000 car 2548 departs Harrington Road with a Beckenham Junction–Wimbledon service on 18 March 2022. The entrance to Beckenham Crematorium and Cemetery can be seen on the left. *Alan Yearsley*

Above: This night shot of Birkbeck shows CR4000 car 2540 forming a Beckenham Junction service on the evening of 16 April 2022. In the foreground on the right can be seen the heavy rail track and platform. *Robert Pritchard*

Chapter 3: London Trams Route By Route

Left: Variotram 2560 pauses at Beckenham Road with a Beckenham Junction–Wimbledon service on 18 April 2022. The heavy rail line can be seen on the right. *Robert Pritchard*

Below: CR4000 car 2543 leaves the terminus at Beckenham Junction with a Wimbledon service on 26 March 2022 as Class 455 EMU 455805 (coupled to 455833 out of sight at the rear) waits to depart with the 15.15 to London Bridge. This scene is now unrepeatable, as the Southern Class 455s were all withdrawn with the May 2022 timetable change and their duties taken over by Class 377 Electrostars. *Brian Garvin*

Shortly after leaving Beckenham Road, the Chatham Main Line between London Victoria and the Medway Towns curves towards us on the left, and is separated from the tramway by some well-established allotments. These allotments are bounded on the third side by the embankment that carried the chord on which London Chatham & Dover Railway Victoria–Kent House–Crystal Palace Low Level/Norwood Junction services sporadically operated until the early 20th century. Most of the chord survived as a siding from Kent House station until 1928.

Here we may see a Southeastern train formed of a Class 465 Networker EMU of 1990s vintage on a suburban service or a longer-distance train to or from Dover or Ramsgate formed of a Class 375 Electrostar unit. As the Chatham Main Line joins us and runs alongside the tramway and the Crystal Palace–Beckenham Junction line, we pass over the Lewisham–Hayes line also known as the Mid Kent Line (despite the entire length of this line being within Greater London, albeit with the southern part of it running through the Borough of Bromley, most of which was historically part of the county of Kent). We then pass over the River Beck, from which the town of Beckenham gets its name.

Just before we arrive at the terminus another railway line curves in from the left, this time a spur line linking the Mid Kent and Chatham Main Lines, which is used for empty stock movements and occasionally by diverted trains. At the time of writing there was

CROYDON: TRAM TO TRAMLINK

also one regular service train booked to use this spur to keep up train crew route knowledge: the 00.20 Cannon Street–Tunbridge Wells. This train was booked to pass Beckenham Junction at 00.40 and we are thus unlikely to see it when exploring the tram network! This spur was also used by coal trains from Bricklayers Arms to the Beckenham Junction coal yard on the far side of the station (now occupied by a Waitrose supermarket) until the mid-1980s. The A2015 Rectory Road then curves in from the right, and we run alongside this road for the last few metres of the route.

Beckenham Junction tram terminus consists of an island platform located just outside the heavy rail station. At times it is possible to see two trams stood here at the same time. A small space housing one of the station's car parks separates the tram terminus from the station, and the station forecourt is immediately adjacent to the tramway buffer stops giving easy interchange between light and heavy rail.

CENTRAL CROYDON LOOP

Croydon town centre is served by a loop that operates in one direction only. Trams that start and finish at New Addington take a complete circuit of the loop. Conversely, it is impossible for a tram from Wimbledon to travel all the way around the loop without reversing because there is no triangular junction at the point where the eastbound and westbound tracks diverge just west of East Croydon station. This can be an inconvenience at times of disruption, and the proposed (but now shelved) Dingwall loop could have remedied this situation.

Above: Car 2536 is seen after arrival at Beckenham Junction on 18 April 2022. In the background can be seen the station building with the car park and cycle racks in front. *Robert Pritchard*

Above: Car 2541 heads away from East Croydon station along George Street past the junction with Wellesley Road on 17 April 2022. The destination reads West Croydon so this is probably a New Addington–Croydon service. The eastbound track trails in from Wellesley Road just behind the tram. *Robert Pritchard*

Leaving East Croydon and travelling westwards this time, we head along George Street, initially on reserved track as far as the junction with Wellesley Road on a major road intersection where the eastbound track trails in on our right. The westbound track continues along the remaining part of George Street and runs on-street from this point although only trams, pedestrians, cyclists and delivery vehicles are permitted on this section. Shortly before reaching the end of this narrow shopping street we arrive at its namesake stop, which consists of a single platform on our right, almost directly outside the Wetherspoon's pub "The George" and opposite a Tesco store. This is almost immediately followed by the Whitgift Almshouses, and we then continue downhill through a pedestrianised shopping area as George Street becomes Church Street, after which the next stop is named. The tower of Croydon Minster can be glimpsed to the left after leaving Church Street.

After Church Street stop, the tram runs onto a short section of interlaced track that continues as far as the junction with Tamworth Road, where Wimbledon-bound trams continue straight on. This is the first of two stretches of interlaced track on the network (see also under Wimbledon). Here, trams taking a full circuit of the loop turn right onto a triangle junction, the other half of which is used by trams coming from Wimbledon. We then continue along the left-hand side of Tamworth Road (lines running on-street at one side being commonly referred to as "gutter running" in tramway parlance) to the next stop, Centrale, which opened in 2005 to serve the namesake shopping centre on our right. The impressive Tamworth Halls building is to the left, this incorporates a day nursery and healthcare centre. To date Centrale is the only stop on the entire network not

Chapter 3: London Trams Route By Route

Above: Car 2538 leaves George Street stop with a New Addington service on 2 September 2022. This tram will take a full circuit of the Croydon town centre loop before heading for its destination. Note that unlike the previous photo the destination indicator is already showing New Addington. *Alan Yearsley*

Above: Car 2536 passes the Whitgift Almshouses towards Church Street with a Wimbledon service on 17 April 2022. *Robert Pritchard*

CROYDON: TRAM TO TRAMLINK

Right: Car 2564 runs downhill through the pedestrianised shopping area between George Street and Church Street stops with a Wimbledon service on 5 April 2019. This and the previous photo show how easily trams fit into pedestrian precincts. *Robert Pritchard*

to have opened at the same time as services started in 2000, apart from a temporary stop set up at Dingwall Road during track renewal work at East Croydon in 2009. The line then carries on along the edge of Tamworth Road, which becomes Station Road as we cross over North End, where the entrance to West Croydon station is located, then arrive at the tram stop of the same name. Anyone interchanging between train and tram here has just a short walk along Station Road. Alternatively, the station has a secondary entrance leading directly onto Platform 4 from Station Road just north of the tram stop for passengers already holding valid tickets, Oyster cards or contactless payment cards or devices. The tram stop is also located directly opposite the important West Croydon bus station for ease of interchange between bus and tram.

Shortly after passing West Croydon bus and rail stations, the line curves to the right and heads in a north-easterly direction along the last section of Station Road, then takes a sharp curve onto the A212 Wellesley Road, a street dominated by high-rise apartment and office blocks. Here the line runs down the centre of the dual carriageway until we pass the entrance

Above: Car 2560 calls at Church Street with a Wimbledon service on 17 April 2022. *Robert Pritchard*

Left: Car 2550 is seen on Tamworth Road shortly after passing over the triangle junction at the end of Church Street on 7 August 2018. The destination on the side indicator reads West Croydon, which suggests that this is a New Addington–Croydon service. *Robert Pritchard*

Chapter 3: London Trams Route By Route

Left: Car 2535 heads away from the camera towards West Croydon with an Elmers End service on 17 April 2022. Centrale shopping centre can be seen behind the tram, and its namesake tram stop is just visible in the background on the left. *Robert Pritchard*

to the Whitgift Centre, Croydon's other main town centre shopping complex, on the right. A pedestrian subway runs underneath the road near the Wellesley Road stop. The line then crosses over the A212 road and runs along the left-hand carriageway as it passes the last stop on the loop, Wellesley Road. Soon after leaving this stop, the loop rejoins the westbound track at the crossroads with George Street and then arrives back at East Croydon.

Right: Car 2543 heads away from the camera as it passes the main entrance to West Croydon station towards the adjacent tram stop on a New Addington–Croydon service on 7 August 2018. *Robert Pritchard*

Below: A busy scene at West Croydon on 13 May 2022 as passengers wait to board the New Addington-bound car 2534. Alongside it is a 264 bus in the adjacent bus station bound for Tooting; this route follows part of the erstwhile trolleybus route 630 to Harlesden via Mitcham Common, Tooting and Putney. *Alan Yearsley*

CROYDON: TRAM TO TRAMLINK

Above: Car 2536 crosses over the A212 Wellesley Road as it passes the entrance to the Whitgift shopping centre and heads towards East Croydon on 17 April 2022. *Robert Pritchard*

Above: Car 2535 is seen shortly after leaving Wellesley Road stop as it heads towards the junction with George Street with an Elmers End service on 17 April 2022. On the left is a 403 bus bound for Warlingham, an area also served by Whyteleafe and Whyteleafe South stations on the Caterham branch (which was mooted for conversion to light rail in the 1980s) and Upper Warlingham station on the East Grinstead line. *Robert Pritchard*

Chapter 3: London Trams Route By Route

Above: Car 2537 arrives at East Croydon with a New Addington service on 11 August 2018 having completed a full circuit of the Croydon town centre loop. *Robert Pritchard*

WIMBLEDON

Continuing straight on at the junction of Church Street and Tamworth Road this time, as the other half of the triangle junction with the town centre loop trails in on our right our line curves to the right onto Cairo New Road and passes Reeves Corner stop, the only stop on the network with tracks running in both directions but served in one direction only. The presence of a flyover

Above: Car 2555 rounds the curve near Reeves Corner with a Wimbledon service (right), while on the left car 2550 heads away from the camera towards central Croydon with a Wimbledon–Beckenham Junction service on 17 April 2022. *Robert Pritchard*

CROYDON: TRAM TO TRAMLINK

Above: Car 2550 calls at the unidirectional Reeves Corner stop with a Wimbledon–Beckenham Junction service on 17 April 2022. On the left is the westbound track where Wimbledon-bound trams pass non-stop. *Robert Pritchard*

Above: Car 2546 calls at Wandle Park with a Wimbledon–Beckenham Junction service on 2 September 2022. *Alan Yearsley*

Chapter 3: London Trams Route By Route

Above: Car 2549 passes the tower of Croydon A Power Station as it heads away from Waddon Marsh stop with a Wimbledon–Elmers End service on 13 May 2022. On the right is a footpath that leads to the next stop, Wandle Park. *Alan Yearsley*

Above: Car 2555 departs from Ampere Way with a Wimbledon–Beckenham Junction service on 17 April 2022. On the right are the two former towers of Croydon B Power Station, which now form part of an IKEA furniture store, hence the blue and yellow IKEA logos at the top of each chimney. *Robert Pritchard*

CROYDON: TRAM TO TRAMLINK

Above: Car 2556 departs from Therapia Lane with a Wimbledon–Elmers End service on 15 April 2022. *Robert Pritchard*

Above: Car 2532 heads away from Beddington Lane stop (located just behind the camera) past the site of the original Beddington Lane station with a Wimbledon–Beckenham Junction service in the early evening of 18 March 2022. The footpath on the right leads from the tram stop to its namesake road. *Alan Yearsley*

Chapter 3: London Trams Route By Route

Above: Car 2536 crosses the flyover over the Balham–Sutton line as it approaches Mitcham Junction with a Wimbledon service on 15 April 2022. *Robert Pritchard*

Above: Car 2562 calls at Mitcham Junction with a Wimbledon–Elmers End service on 13 May 2022. A Southern Class 377/6 EMU can be seen in the adjacent heavy rail station on the right. *Alan Yearsley*

CROYDON: TRAM TO TRAMLINK

Left: Car 2552 traverses the interlaced track as it approaches Mitcham with a Wimbledon–Beckenham Junction service on 17 April 2022.

carrying the A236 Roman Way precluded the provision of a platform for Wimbledon-bound trams without causing inconvenience to passing pedestrians not waiting for trams so only eastbound trams call here. This stop, and its namesake street, gets its name from the nearby House of Reeves furniture shop, which has been in existence since 1867. This was the site of some of the infamous riots in London during August 2011, when part of the furniture store was set alight.

After passing Reeves Corner, the line takes a sharp curve to the left and passes underneath this flyover, then reduces to a single track and runs briefly alongside the West Croydon–Sutton line until Cairo New Road becomes Waddon New Road. Here, the line climbs onto a flyover and passes over the heavy rail tracks at the site where the Wimbledon line formerly diverged from the Sutton line before the former was

Below: Car 2557 departs from the island platform stop at Phipps Bridge with a Wimbledon–Beckenham Junction service on 18 April 2022. *Robert Pritchard (2)*

Chapter 3: London Trams Route By Route

Above: Car 2552 passes an industrial warehouse and office complex between Morden Road and Phipps Bridge with a Wimbledon–Beckenham Junction service on 18 April 2022. *Robert Pritchard*

Above: Car 2565 pauses at Merton Park with a Wimbledon service in the early evening sunshine on 13 May 2022. *Alan Yearsley*

CROYDON: TRAM TO TRAMLINK

Above: A group of pedestrians and cyclists wait to cross the track at Dundonald Road as car 2556 arrives with a Wimbledon service on 13 May 2022. *Alan Yearsley*

Above: Car 2536 calls at Dundonald Road with a Wimbledon–Elmers End service on 13 May 2022. *Alan Yearsley*

Chapter 3: London Trams Route By Route

Above: Car 2544 stands in Platform 10a at Wimbledon after arriving with a Route 3 service on 2 December 2015, shortly after the opening of the new Platform 10b which can be seen on the left. *Robert Pritchard*

converted to light rail. We then skirt around the edge of Wandle Park, and the line becomes double track as it descends from the flyover towards the park's namesake stop. There has been a lot of new development here, with new flats and housing to the left hand side. A feature of the Wimbledon line – a footpath alongside – can be seen on the left from this point.

From here the line heads in a north-westerly direction as it passes through an area dominated by retail parks and industrial estates. The tower of Croydon A Power Station, now owned by Whitetower Energy Company, can be seen on our right just before we arrive at the next stop, Waddon Marsh, where Turners Way Gas Works were located immediately adjacent to the stop until they were demolished in January 2022. The footpath alongside the line switches to the right hand side at this point. We then pass beneath the A23 Purley Way and the two former towers of Croydon B Power Station are seen on our left. These now bear the blue and yellow IKEA logo and form part of an IKEA furniture store, the car park for which is located adjacent to Ampere Way stop so we may see a few passengers alighting bound for IKEA or boarding carrying packs of flat pack furniture from here!

For the next few miles, the line passes through a mixture of industrial and residential areas. Just beyond the next stop, Therapia Lane, the tram depot can be seen on our left where we may be able to spot a few trams that are stabled there. There is also a crossover here and adjacent to the depot there is a staff halt with a westbound platform on our left followed by the eastbound platform on the right, although at the time of writing this halt had not been in regular use since the Covid pandemic because its short narrow platform prevented social distancing and it is located less than 80 m west of the passenger stop. The sidings just beyond the depot continue until we are about two-thirds of the way towards the next stop, Beddington Lane (there is a second crossover the other side of the depot). Just before arriving here, the line crosses over the B272 road from which this stop gets its name, with the tram stop being sited immediately to the west of the site of the original station which was located adjacent to the level crossing. The stop here can sometimes echo to the sounds of speed bikes on the adjacent course!

On a very straight section we pass a golf course to the left which includes a foot crossing of the tramway. Beddington Lane Industrial Estate is to the right, and it feels semi-rural as far as the next stop, Mitcham Junction. Before arriving here the line climbs onto a single track flyover over the Balham–Sutton line, which then runs alongside the tramway with the tram stop being located alongside the heavy rail station. This single track section can be a particular pinchpoint of the network following the increase in frequency on the Wimbledon line. The tram and railway lines continue to run side by side as they pass under the A237 Carshalton Road, with the tram line being reduced to single track beneath the road bridge after which the railway curves off to the right as the tramway continues straight ahead and once again becomes double track straight away. The area through which we pass becomes more built-up as we approach the next stop, Mitcham, and this is immediately followed by the second set of interlaced track as the line passes beneath the A217 London Road. As the line reverts to a double track formation the A239 Morden Road runs alongside us on our left for a short distance, then curves off to the left as residential areas give way to an industrial estate that continues as far as the next stop, Belgrave Walk. This is the first of two island platform stops on the Wimbledon line, after which we run through a mixture of open fields and housing developments. The next stop, Phipps Bridge, also has an island platform and gets its name from the adjacent council estate on our right, while on the left is Morden Hall Park, a public park owned by the National Trust. The island platforms at these two stops are both quite wide and generous.

Shortly after leaving Phipps Bridge, the last short single track section starts, and we soon cross over the River Wandle and then run alongside Deer Park Road at the edge of another industrial estate

CROYDON: TRAM TO TRAMLINK

on our right. The single track gives way to double track as the line passes under the A24 Morden Road and arrives at its namesake stop. After leaving Morden Road, we pass Nursery Road Playing Field on our right, where it may be possible to make out the remains of the trackbed of the former Merton Abbey line to Tooting at the edge of the field as we approach the penultimate intermediate stop, Merton Park, now again with houses on either side.

The line then takes a sweeping curve through the southern outskirts of Wimbledon as it crosses over Dundonald Road on a flat crossing and passes its namesake stop, after which the South Western Main Line trails in on our left as we arrive at the terminus inside Wimbledon station, returning briefly to single line before arriving at the station. A variety of trains can be seen here, including (at the time of writing) Govia Thameslink Class 700 EMUs on Sutton Loop services and South Western Railway Class 159 DMUs and Class 444, 450 and 455 EMUs (although the 455s are now on borrowed time pending the introduction of the new Class 701s). Trams arrive at the far right-hand side of the station as you look towards central London. Originally only one platform was provided for trams and was created using one half of the island platform used by Sutton Loop services. In 2015 the tram terminus at Wimbledon was expanded to create a second platform numbered 10b with the existing tram platform then being numbered 10a. Both platforms are used in normal service.

Below: The headshunt at the end of Platform 10b. In the background on the left is Wimbledon signal box. *Robert Pritchard*

LONGEST DISTANCES BETWEEN STOPS (ALL OVER 1 KM LISTED)

Sandilands–Lloyd Park	1.60 km
Lloyd Park–Coombe Lane	1.58 km
Coombe Lane–Gravel Hill	1.40 km
Beckenham Road–Beckenham Junction	1.23 km
Arena–Harrington Road	1.12 km
Mitcham–Mitcham Junction	1.12 km
Mitcham Junction–Beddington Lane	1.07 km
Beddington Lane–Therapia Lane	1.06 km

SHORTEST DISTANCES BETWEEN STOPS

Centrale–West Croydon	0.30 km
Phipps Bridge–Belgrave Walk	0.38 km
Reeves Corner–Centrale	0.39 km
George Street–Church Street	0.39 km
Wellesley Road–East Croydon	0.44 km
East Croydon–Church Street	0.46 km
Wimbledon–Dundonald Road	0.47 km

CHAPTER 4:
THE TRAM FLEET

Above: CR4000 car 2532 arrives at East Croydon with an Elmers End service on 8 February 2003, in the original Tramlink livery and in as-built condition with manually operated roller destination blinds. Note the logo of train operator SouthCentral (since rebranded as Southern) and the faded Connex logo on a yellow background on the shop front just visible on the right, Connex having surrendered the SouthCentral franchise to Govia in August 2001. *Robert Pritchard*

The original tram fleet consists of 24 Bombardier Flexity three-section cars, built by Bombardier in Wien (Vienna), Austria, in 1998–99, of which 23 are in service at the time of writing. These are also known as CR4000s because they are closely based on the K4000 fleet in Köln (Cologne), Germany, although the CR4000 version is 1.5 m longer at 30.2 m. The K4000s also usually run in pairs, whereas the CR4000s (and the newer Variotrams) always operate singly. Between 2011 and 2017 these were joined by 12 Stadler five-section Variotrams, taking the total fleet size up to 36. The CR4000s and K4000s are also sometimes referred to as City Trams.

THE CR4000S

At the time of this book going to press the latest TfL Commissioner report, published in February 2023, indicated that TfL was looking to replace the increasingly unreliable CR4000 fleet with procurement for new vehicles

Right: The CR4000s were based on the K4000s already in operation in Cologne, Germany. At the time of this book going to press the Cologne K4000s were still in service but were due to be replaced by Alstom Citadis trams soon. Here a pair of K4000s is seen in Cologne city centre with a Line 7 service to Zündorf on 1 June 2009. *Keith Fender*

CROYDON: TRAM TO TRAMLINK

Above: CR4000 2531 at Mitcham with a Wimbledon service on 8 February 2003. Apart from the all-over advertising liveries, a number of trams carried these "Learn Direct" posters at this time. *Robert Pritchard*

passengers assuming a load of four passengers per square metre. 76% of each vehicle is low-floor, with a short high-floor section at each outer end over the motor bogies.

Each car has eight pairs of sliding plug doors, two on each side of each driving section. There is also an access door on the right-hand side of each driving cab but not on the left because the space on the left-hand side of the cab is taken up by the controls. This design feature has been copied from the K4000s on which the CR4000s are based and means that the external cab access door is on the same side as the pavement on the K4000s but on the off-side on the CR4000s because the UK drives on the left whereas Germany, like the rest of mainland Europe, drives on the right. The external cab door is also on the off-side at tram stops with a conventional two-platform layout but on the correct side at island platform stops and at some termini.

expected to start in late 2023. Full funding for their replacement had not yet started, however, and it is expected that TfL will lease any new vehicles rather than purchasing them outright unless the DfT can provide funding for such a purchase. In the spring of 2023 the CR4000s were suffering from poor reliability leading to ad-hoc service cancellations. By this time the CR4000 fleet was between 24 and 25 years old, and it should be remembered that the oldest Manchester Metrolink T68s were almost as old when withdrawn in 2014 with the West Midlands Metro T69s having only managed 16 years (although by contrast the Sheffield Siemens-Duewag Supertram fleet is now 30 years old and still operating reliably, with no immediate plans for replacement).

The CR4000s use a numbering scheme that carries on from where the old London tram fleet left off and are thus numbered 2530–2553, the highest numbered first generation London tram having been 2529.

In terms of the interior, the CR4000s have 70 seats almost all of which are in a 2+2 layout in facing bays of four although there are two single longitudinal seats, two pairs of facing longitudinal seats in the short centre section, and two wheelchair spaces. There are also two pairs of seats facing the driving cab at each end so that passengers seated at the front end can enjoy a driver's eye view. The CR4000s are also designed to carry up to 138 standing

Above: An original CR4000 interior taken in 2003. *Robert Pritchard*

Right: The driver enters the cab via the side door on car 2549 at Wimbledon on a Beckenham Junction service on a wet 11 June 2023 after alighting to change the points manually due to a points failure. After arriving and setting down the passengers at Platform 10b, the tram then proceeded as far as the points, where the driver left and then re-entered the cab. *Robert Pritchard*

Chapter 4: The Tram Fleet

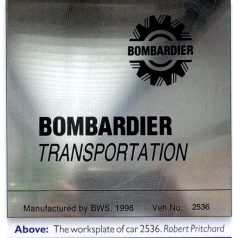

Above: The worksplate of car 2536. *Robert Pritchard*

Above: Still in the original livery but with digital destination displays, car 2548 passes the Cricketers pub as it departs Lebanon Road with an Elmers End service on 7 October 2008. *Keith Fender*

Right: Interior of car 2536 at Beckenham Junction on 18 April 2022. Note the two different versions of seat moquette fitted on refurbishment, with the second variant being visible in the background. It is loosely based on the upholstery used on some London buses and Underground trains in the 1970s and 1980s. *Robert Pritchard*

Above: The area adjacent to one of the entrance doors on car 2536. Note the perch seats next to the door. *Robert Pritchard*

Right: Interior of the cab of CR4000 2546 at Wimbledon on 11 June 2023. *Robert Pritchard*

CROYDON: TRAM TO TRAMLINK

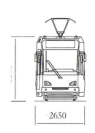
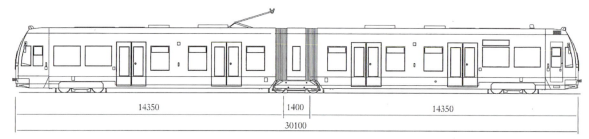
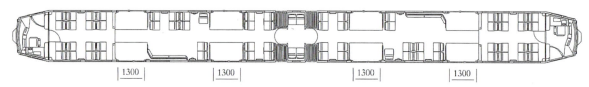

Above: Line drawing of the Bombardier CR4000 for Croydon. *Courtesy Bombardier*

Right: From October 2008 onwards the original red/white/black livery started to give way to a new green/white/blue colour scheme. 2549 sports the new livery as it arrives at East Croydon with a New Addington service on 10 February 2009. *Robert Pritchard*

Where the cab door is on the off-side drivers can still use it if they take the necessary precautions to avoid being hit by oncoming trams, and if there is sufficient space for them to board and alight, as there are step apertures on the bodyside beneath the cab door. However, experience has shown that they often tend to use the internal cab door to enter and leave via the passenger saloon where this is the case. During driver changes at East Croydon you may see the relief driver enter the cab via the external door and the outgoing driver leave via the saloon or vice versa. Driver changes mostly take place here or at Therapia Lane, which is a conventional two-platform stop. At Elmers End and Wimbledon, the cab door will be on the off-side on arrival and on the correct side at the departure end of the platform. At Beckenham Junction and New Addington, this will depend on which side of the island platform is used. The Variotrams do not have external cab doors but do have a passenger door at each end immediately adjacent to the cab.

CR4000 MODIFICATIONS

In 2006 the traditional roller blind destination indicators on the CR4000s were replaced by digital destination displays. At the same time the on-board announcements were updated, featuring the voice of BBC broadcaster and journalist Nicholas Owen. Then between 2008 and 2009 the CR4000 fleet received its first refurbishment, with the original dark blue seat upholstery being replaced by a red, grey and two-tone green chequered moquette in a similar pattern to those used on some London buses and Underground trains in the 1970s and 1980s. This was followed by a second refurbishment between 2015 and 2016 involving a full interior repaint, new flooring, new safety warning signs, repainting of the interior handrails light green instead of yellow, and removal of the stop request buttons, request stops having been abandoned in favour of all trams automatically calling at all stops by this time. Also as part of this refurbishment programme LED headlights were fitted.

So far only one CR4000 tram has been named. 2535 was named Stephen Parascandolo 1980–2007 in a ceremony at Beckenham Junction on 20 October 2007 in honour of Stephen Parascandolo, a local tram enthusiast and webmaster of the unofficial Croydon

TABLE 1: CR4000 TECHNICAL DATA

Built:	1998–99 by Bombardier, Vienna, Austria
Number series:	2530–2553
Wheel arrangement:	Bo-2-Bo
Traction motors:	4 x 120 kW (161 hp)
Line voltage:	750 V DC
Track gauge:	1435 mm
Seats:	70
Total passenger capacity (4 pass/m^2):	138
Weight:	36.3 tonnes
Dimensions:	30.1 x 2.65 m
Braking:	Disc, regenerative and magnetic track
Couplers:	Scharfenberg
Maximum speed:	80 km/h (50 mph)

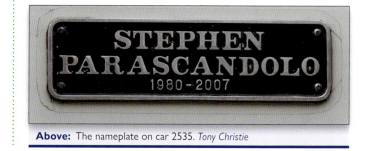

Above: The nameplate on car 2535. *Tony Christie*

Tramlink website who was killed in a road accident at the age of 26, since when that website has remained live but dormant as no-one has ever been found to take it over.

CR4000 LIVERIES

The CR4000s have carried two different standard fleet liveries to date. Until their first refurbishment in 2008 they carried the then standard Tramlink colour scheme of red and white with black window surrounds. This was based on the livery latterly carried by London's first generation trams, and was also not dissimilar to that of the Cologne K4000s in the late 1990s. They also featured gold cab front numbers in the same style as those carried on London's original trams from the 1930s onwards. The post-2008 livery retains the same style of numerals but they are now dark blue.

Car 2550 was initially outshopped in First Group corporate livery while four others carried advertising liveries: 2531 for Addington Palace Country Club, 2533 for Nescafé, Nestlé UK's headquarters then being located in Croydon, 2542 for Amey Construction, and 2546 for the Whitgift Shopping Centre in central Croydon. By early 2003 all trams carried the standard livery, and with the first refurbishment in 2008 this was replaced by a new standard livery of light green, white and blue to coincide with the change of ownership to TfL as mentioned in chapter 2. Since then a number of other advertising liveries have been applied to CR4000s (see Table 3).

Above: 2550 was delivered in First Group livery to reflect the company's involvement in Tramlink. It is seen at East Croydon on a Route 2 service shortly after the start of passenger services in 2000. *John Law*

Above: 2546 carried an all over advert for the Whitgift shopping centre in the early days of Tramlink. It is seen heading away from the camera between Lebanon Road and East Croydon with a New Addington service. *Geoffrey Skelsey*

Above: 2531 in an earlier version of the McMillan Williams Solicitors advertising livery at East Croydon on 24 October 2014. *Keith Fender*

CROYDON: TRAM TO TRAMLINK

Above: 2531 is seen again, this time just east of East Croydon with a New Addington service in the second version of McMillan Williams livery carried by this tram on 5 July 2021. Fellow CR4000 2534 carried yet another advertising livery for the same company, consisting of red beneath the windows and white above. *Keith Fender*

Above: 2550 arrives at East Croydon on a Beckenham Junction service in a special Tramlink safety yellow livery on 15 October 2016. At least four other trams have also carried all over yellow adverts: 2532, 2539 and 2553 for IKEA (albeit with the cab ends retaining the standard livery) and 2542 for Amey Construction. *Robert Pritchard*

Chapter 4: The Tram Fleet

Above: 2542 leaves East Croydon in the earlier all over red version of Turkish Airlines livery on 21 January 2015. As red is the colour that has long been associated with public transport in Greater London, one could perhaps almost be forgiven for thinking that this was the new standard tram livery! *Keith Fender*

Above: 2537 calls at Mitcham in the later version of Turkish Airlines livery on 17 March 2017. *Tony Christie*

CROYDON: TRAM TO TRAMLINK

Above: As well as all over adverts, the trams have also often been adorned with poppies on the sides or cab fronts in the run-up to Remembrance Day. Here car 2549 calls at George Street with a Wimbledon service on 11 November 2013. *Keith Fender*

Above: This photo shows the contrasting cab end designs of the two tram types at Coombe Lane on 17 April 2022. On the left Variotram 2556 forms a Croydon–New Addington service, while on the right CR4000 2546 has just arrived from New Addington and is bound for central Croydon. Note the lack of wing mirrors on the Variotram unlike the CR4000. *Robert Pritchard*

Chapter 4: The Tram Fleet

THE STADLER VARIOTRAMS

London Trams' 12-strong Stadler Variotram fleet was built by Stadler Pankow in Berlin, Germany, and based on those already in operation in Bergen, Norway. Other tramways with versions of the Variotram (known as the Variobahn) include the Austrian city of Graz and the German cities of Bochum/Gelsenkirchen, Chemnitz, Mannheim/Ludwigshafen, München (Munich), Nürnberg (Nuremberg) and Potsdam. Variotrams also operated in Sydney, Australia, from 1997 until 2015. Since its introduction in 1993 the Variotram has been built successively by ABB, Adtranz, Bombardier, and since 2001 by Stadler.

Unlike the CR4000, the Variotram is 100% low-floor and has four more seats but with space for four less standing passengers (134 rather than 138). There are also two areas with a wheelchair space and six perch type seats each, located in the second and fourth section of the tram. These spaces can also be used by passengers with pushchairs, shopping trolleys or bulky items of luggage, for example. The ordinary seats in these sections are unidirectional, facing the same way as the adjacent driving sections, while the seats in the centre section and at the outer ends are mainly in facing bays but with a single seat each side of the aisle facing the cab and adjacent to the articulation in the centre section. There is also a pair of seats each side of the aisle facing the articulation at the inner end of each driving section.

As with the CR4000s, the Variotrams also have eight pairs of sliding plug doors, but with one pair of doors on each side of each of the four sections and again none in the short centre section. Unlike the CR4000s, the Variotrams have no external cab doors but the passenger doors at each outer end are immediately adjacent to the cab. Another notable feature of the Variotrams is that they do not have wing mirrors, with the rear facing cameras being used instead.

Although the CR4000 fleet was still operating reliably, by 2011 an additional batch of trams was needed to cope with a 45% increase in traffic since the network opened. Initially TfL hoped to lease vehicles, with sources confirming to the media that a deal to lease some of Edinburgh's under-employed fleet of CAF vehicles was the obvious solution. However, Edinburgh offered TfL some of its trams but not at the level of price that TfL was looking to pay (and they were also too long so would have required modifications to the platforms) so a fleet of new trams would be needed instead.

Early in 2011 the three shortlisted suppliers for the new trams were announced as CAF, Stadler Pankow and Polish builder PESA. In August of that year London Mayor Boris Johnson announced that Stadler had won the contract to supply six Variobahn trams with an option for up to eight more. Three of the Variotrams for Tramlink were diverted from a batch of five built for Bergen. This meant that Stadler could produce the necessary trams within the required timescale, which might not have been the case if a completely new-build fleet of trams had been ordered. Stadler could also demonstrate that purchasing outright offered better value for money than leasing.

Minor modifications were made to the Variotrams originally built for Bergen to comply with UK regulations. These included a redesign of the cab ends to meet UK crashworthiness standards, addition of UK-compliant bells and horns, and fitting of new lights including fog lights which are required in Croydon but, strangely, not in Norway. The interior layout is also different from the Bergen Variotrams. Only minimal alterations to the existing tramway infrastructure were needed, including making platforms at some tram stops slightly narrower using an angle grinder but the platforms did not need to be extended.

The first Variotram, 2554, was delivered to Therapia Lane depot in the early hours of 9 January 2012 after being tested on the Chemnitz tramway in eastern Germany. 2554 was launched to the specialist and local media on 15 February. 7 June that year saw the arrival of car 2559, the last of the initial Variotram order. The new trams were initially used on peak hour extras, with 2555 being the first to run on a regular timetabled (as opposed to an additional) service on 29 May.

In 2013, as part of its ten-year business plan, TfL announced that it would exercise an option for four more new trams, and on 4 May that year the first of the four additional Variotrams, 2560, was delivered, then in 2016 two additional Variotrams were ordered taking the total number to 12.

VARIOTRAM LIVERIES

All Tramlink Variotrams were delivered in the by then standard green, blue and white Tramlink livery which they still carry today. At the time of writing only one Variotram had carried an advertising livery: 2554 received an all over purple "Love Croydon" livery in June 2012, just a few months after being delivered, and still carries this advert to this day. In addition, Variotrams 2560 and 2562 were delivered in a special "New trams for Croydon" livery on one end only, covering just under half of the tram with the rest of the vehicle carrying the standard Tramlink livery. This consisted of a pale green background bearing a mixture of vertical and horizontal blue and grey TfL "roundel" logos. Both vehicles entered service in late 2015 but have since lost this special colour scheme and now carry the standard livery throughout.

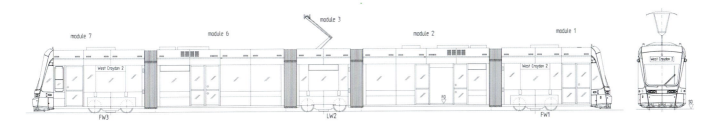

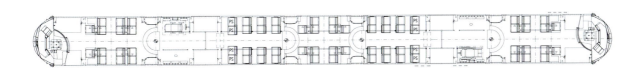

Line drawing of the Variotram for Croydon. *Courtesy Stadler*

CROYDON: TRAM TO TRAMLINK

TABLE 2: VARIOTRAM TECHNICAL DATA

Built:	2011–16 by Stadler, Berlin, Germany
Number series:	2554–2565
Wheel arrangement:	Bo-2-Bo
Traction motors:	8 x 45 kW (60 hp)
Line voltage:	750 V DC
Track gauge:	1435 mm
Seats:	74
Total passenger capacity (4 pass/m^2):	134
Weight:	41.5 tonnes
Dimensions:	32.4 x 2.65 m
Braking:	Disc, regenerative and magnetic track
Couplers:	Albert
Maximum speed:	80 km/h (50 mph)

The following table on page 64 shows all advertising and other non-standard liveries carried by the Tramlink fleet to date, to the best of our knowledge.

Left: The cab layout at the "A" end of Variotram 2554. *Tony Christie*

Above: A close-up view of the assistance buttons in the disabled priority area on a Variotram, including an intercom to enable passengers to speak to the driver, an emergency alarm and a stop request button (although the latter is now effectively redundant as all trams now normally call at all stops). *Keith Fender*

Above: Interior of Variotram 2554 at Elmers End on 21 March 2013, with bays of four facing seats in the centre section in the foreground and unidirectional seats in the adjacent section. Beyond these can be seen the driving section and the cab in the background. *Robert Pritchard*

Above: The builder's plate of Variotram 2554. *Keith Fender*

Chapter 4: The Tram Fleet

Above: Variotram 2554 on test in Chemnitz, Germany, on 3 January 2012, showing the somewhat optimistic destination of Crystal Palace! *Keith Fender*

Above: Variotram 2560 is seen on a low loader in a layby in Whyteleafe on its way to Therapia Lane depot on 3 May 2015. The yellow banner in the cab window reads "convoi exceptionnel" (literally exceptional shipment, i.e. wide, long or dangerous load). *Keith Fender*

CROYDON: TRAM TO TRAMLINK

TABLE 3: ADVERTISING AND NON-STANDARD LIVERIES CARRIED TO DATE

Car	Advert	Colour(s)	Dates carried
2531	Addington Palace	Purple	15/16 Jan 2000–?
2531	McMillan Williams Solicitors	Black with white cab ends	Sept 2014–Nov 2015
2531	Taylor Rose MW	Pink	July 2021–July 2022?
2532	IKEA*	Yellow	March 2004–?
2533	Nescafé	Red/black with white (originally red) doors	4/5 Sept 1999–?
2534	McMillan Williams Solicitors	Red/white	May 2013–Dec 2015
2537	Turkish Airlines*	Red/blue/black/green	Aug 2016–?
2539	IKEA*	Yellow	March 2004–?
2542	Amey Construction	Yellow	8/9 Oct 1999–Oct 2001
2542	Turkish Airlines	Red	Dec 2014–Feb 2016
2546	Whitgift Shopping Centre	Blue	Nov 1999–?
2548	The Simpsons Movie	Yellow with pink doors	Late 2007–Early 2008
2548	Walls Magnum Infinity ice cream	Pink	April–May 2012
2550	First Group	Dark blue/pink/white	April 2000–Oct 2001
2550	Tramlink Safety	Yellow	Jan 2015–Nov 2016
2551	Energy Efficiency#	Dark blue	March 2005–?
2552	Grants entertainment venue	Various (mainly black/pink)	Early 2007–Early 2008
2553	IKEA*	Yellow	March 2004–?
2554	Love Croydon	Pink/purple	June 2012–present
2560	New Trams for Croydon$	Pale green with TfL "roundel" logos	Nov 2015–?
2562	New Trams for Croydon$	Pale green with TfL "roundel" logos	Dec 2015–?

\# One side of one section only

$ One end of the vehicle only

* These liveries retained the standard fleet livery on the cab ends

Right: Variotram 2554 was outshopped in a special "Love Croydon" livery in June 2012 shortly after entering service and has carried this all over advert ever since. On 18 April 2022 it heads towards the flyover over the West Croydon–Sutton line with a Wimbledon–Beckenham Junction service. *Robert Pritchard*

Left: Variotrams 2560 and 2562 were delivered in a special "New Trams for Croydon" livery on one end only. Here 2560 departs from Merton Park with a Wimbledon–Elmers End service on 20 April 2016. *Tony Christie*

Chapter 5: The Depot and Infrastructure

CHAPTER 5:
THE DEPOT AND INFRASTRUCTURE

Most of the network is double track, although there are single track sections in places particularly on the easternmost sections of the Beckenham Junction branch, between Arena (where the Elmers End branch diverges from the Beckenham Junction line) and Elmers End, and between King Henry's Drive (the last intermediate stop on the New Addington branch) and the New Addington terminus. The Croydon town centre loop is also single track, as are parts of the Wimbledon branch west of Beddington Lane although the section between Mitcham Junction and Mitcham was doubled in 2012. There are also short sections of interlaced track just west of Mitcham and immediately west of Church Road stop on the Croydon town centre loop. Between Reeves Corner and Wandle Park and at Mitcham Junction, a single-track flyover was built to carry the tramway over the remaining heavy rail lines.

The network consists of a mixture of on-street running, new-build sections of dedicated right of way, and former heavy rail alignments. However, unlike many other tramways, most of the on-street sections are on tram only or tram and bus only lanes. The section between Park Hill Road, just east of East Croydon, and Chepstow Road, between Lebanon Road and Sandilands stops, is restricted to buses, trams, cyclists, taxis and local traffic (i.e. local residents, workers and visitors to homes and businesses in the neighbourhood) in peak hours so relatively few cars can be seen here. Only the westbound track between the junction with Tamworth Road and Reeves Corner tram stop is shared with all types of motor traffic at all times without restriction.

Stops consist of simple low-level platforms with shelters. There are also CCTV cameras, litter bins, advertisement panels, indicators giving the destination of, and the number of minutes until, the next tram, street maps covering a square mile around each stop, and help points to allow passengers to contact the Tramlink control

Above: The interlaced track on Church Street. The Croydon town centre loop diverges to the left, and the Wimbledon line heads straight on towards the camera. Church Street tram stop can be seen in the background. *Tony Christie*

centre at Therapia Lane depot. Most stops on double track sections have two platforms but a few have island platforms. East Croydon has three platforms, with the platform nearest the heavy rail station (Platform 1) being used by eastbound trams to Beckenham Junction, Elmers End and New Addington and the outer side of the island platform (Platform 3) by westbound trams to

Above: Between Reeves Corner and Wandle Park and at Mitcham Junction, flyovers were built to carry the tramway over the remaining heavy rail lines. Here Variotram 2557 crosses over the West Croydon–Sutton line and heads towards Reeves Corner on a Wimbledon–Beckenham Junction service on 18 April 2022. *Robert Pritchard*

65

CROYDON: TRAM TO TRAMLINK

Above: Tram only lanes are marked by these tram only signs, such as here on George Street with CR4000 2537 seen just west of East Croydon on 5 July 2021. *Keith Fender*

Above: CR4000 2546 calls at Lebanon Road with a New Addington service on 17 April 2022. Alongside the tram is a 119 bus bound for Purley Way. This section is restricted to buses, trams, cyclists, and local traffic to and from homes and businesses in peak hours. *Robert Pritchard*

Chapter 5: The Depot and Infrastructure

Above: This view of Harrington Road stop, with 2536 departing for Beckenham Junction, shows the waiting shelters, platform indicators, litter bins, and the help point affixed to the shelter on the Croydon-bound platform. *Robert Pritchard*

Above & right: Two views of the advertisement panels at Harrington Road, showing the tram network map, local area map, details of how to pay your fare, and times of first and last trams from the stop. Note also the poster prohibiting the carriage of e-scooters and e-unicycles, the signs beneath the stop nameboard banning the consumption of alcohol and smoking (including e-cigarettes), and the sign indicating that trams towards Beckenham leave from this platform. Although too small to read in this view, taken on 17 April 2022, the poster bearing the local area map also still states that face coverings are compulsory on trams (despite TfL having lifted this requirement from 24 February that year, although passengers were still strongly recommended to wear them). *Robert Pritchard (2)*

CROYDON: TRAM TO TRAMLINK

Above & below: Two views of East Croydon tram stop on 30 July 2006 following a fault with a rail joint preventing Platform 1 (out of sight on the right) from being used. This meant that eastbound trams had to use the rarely used Platform 2 instead, leading to concerns among commuters about overcrowding on the island platform. **Above:** 2533 (right) waits to leave Platform 2 with a New Addington service, while 2535 (left) waits in Platform 3 with a westbound service. **Below:** 2533 departs from Platform 2 and heads for New Addington. *Keith Fender (2)*

Chapter 5: The Depot and Infrastructure

Left: At first sight these light signals at Dundonald Road crossing, adjacent to the tram stop of the same name, look like an ordinary set of pedestrian and traffic lights, but they are in fact activated by passing trams. Both the traffic light and the pedestrian signal are green when no tram is due, and these will change to red as a tram approaches from the Croydon direction or is about to depart from the stop. *Alan Yearsley*

West Croydon (including trams arriving from New Addington and taking a full circuit of the Croydon town centre loop) and Wimbledon. Platform 2, the inner side of the island platform opposite Platform 1, is not in regular use but is occasionally used by terminating trams during engineering blockades or other exceptional circumstances such as during the summer of 2006 when a fault with a rail joint prevented eastbound trams from using Platform 1 so Platform 2 had to be used instead. At Elmers End and Wimbledon, the trams use island platforms that are also still used by heavy rail trains on the opposite side of the same platform. Because of this, the level of the tram track has been raised to accommodate the trams rather than lowering the platform level on one side.

In its previous capacity as a heavy rail line, the Wimbledon branch had a number of level crossings, for example at Dundonald Road, Merton Park and Beddington Lane. When the line was converted to light rail the crossing barriers were removed and the crossings (or highway junctions as they are known in light rail terms) are now controlled solely by lights. This is because unlike heavy rail trains, light rail vehicles do not legally have absolute priority over road traffic at a crossing, although the lights are activated by the approaching trams and thus generally act in their favour. There are also highway junctions elsewhere on the network, particularly on the New Addington line, and several foot crossings but these are not usually equipped with lights, only "Tramway: Look Both Ways" signs.

Above: Car 2540 crosses Oaks Road, between Lloyd Park and Coombe Lane, with a New Addington–Croydon service on 17 April 2022. This crossing is likewise equipped with traffic lights activated by approaching trams, and the standard blue "Tramway: Look Both Ways" signs also found at foot crossings, but there are no separate pedestrian signals here unlike at Dundonald Road. *Robert Pritchard*

CROYDON: TRAM TO TRAMLINK

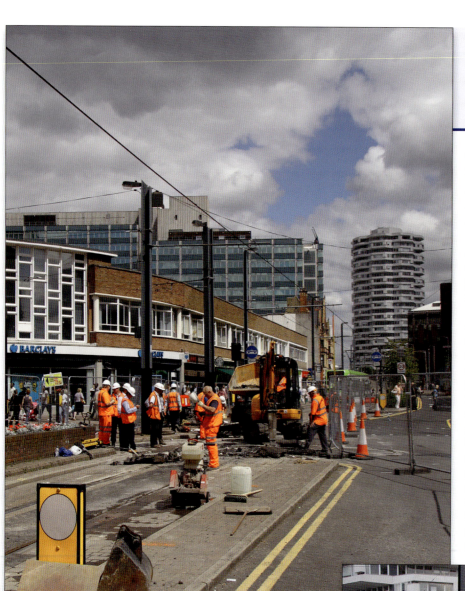

Left: Track renewals in progress at the junction of George Street and Wellesley Road on 3 August 2009, the first day of Phase 2 of that year's blockade. In the background a tram can be seen reversing in the temporary headshunt just west of East Croydon tram stop that this closure had created. *Keith Fender*

TRACK RENEWALS

Between 2009 and 2015 a £54 million upgrade plan funded by TfL saw the replacement of tracks and points along with the resurfacing of the roads across the Croydon town centre loop and installation of a new drainage system at the junction of Cherry Orchard Road and Addiscombe Road just east of East Croydon to prevent flooding during heavy rain.

To enable this work to be carried out, all or part of the loop between Reeves Corner and Sandilands stops was closed during a series of blockades mainly during the Easter and summer holiday periods. These closures meant that the network, and therefore the tram fleet, was physically split, with those trams marooned east of the blockade being unable to return to the depot and having to be stabled overnight in Sandilands Tunnel or on the Elmers End branch with security guards to prevent vandalism. The tunnel could only be used for this purpose at night when trams were not running, so services to Elmers End were suspended in the evenings and on Sundays to allow the branch to be used to stable trams that were not needed.

Three views of track renewal work in progress near East Croydon on 21 April 2011. Old track being removed on Addiscombe Road, at the junction of Cherry Orchard Road just east of East Croydon (above) and immediately east of the station (above right). **Right:** New track awaits installation at the junction of Addiscombe Road and Cherry Orchard Road in this view looking towards Sandilands. *Keith Fender (3)*

Chapter 5: The Depot and Infrastructure

Above: Car 2535 waits to leave Reeves Corner with a Wimbledon service on 15 February 2012 during a blockade of the central Croydon loop, when the stop, normally served only by eastbound trams, was being used as a temporary terminus. Note that as well as a card reader, the ticket machine was then also still in use. These would disappear six years after this photo was taken but were initially covered up with a blue sleeve pending removal. *Tony Christie*

Above: The temporary stop at Dingwall Road on 14 August 2009. Track work can be seen in progress in the background. *Keith Fender*

CROYDON: TRAM TO TRAMLINK

Left: Embedded track renewal in progress at Lebanon Road stop on 13 August 2014, looking west towards East Croydon. *Susan Fender*

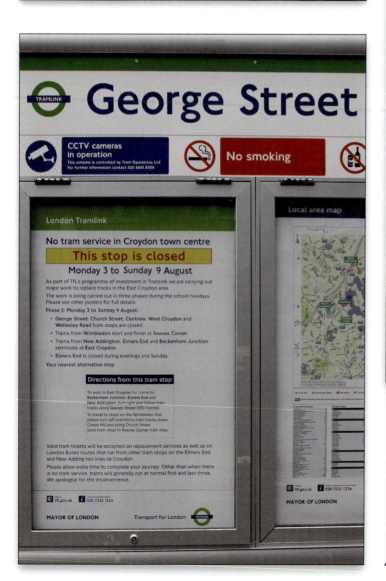

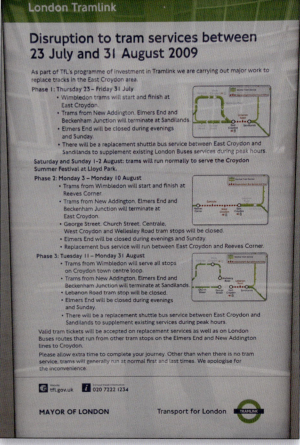

Above & left: Two closure notices displayed in advance of the summer 2009 engineering blockades.

Left: A closure notice at George Street on 3 August, the first day of Phase 2 of the work involving a blockade of the Croydon town centre loop.

Above: A poster giving full details of the closures, seen on 20 July just before the start of Phase 1 of track renewal works. Note that the poster does not mention the temporary Dingwall Road stop, with the map suggesting instead that Wimbledon–Croydon trams will terminate at Wellesley Road and start back from George Street. *Keith Fender (2)*

Chapter 5: The Depot and Infrastructure

Two views of the junction of the Beckenham Junction and Elmers End lines just north of Arena. **Above:** Car 2553 approaches Arena with a service from Elmers End on 18 March 2022. In the foreground is the foot crossing over the Beckenham Junction line. **Below:** An assortment of signs on the approach to the junction, looking towards Elmers End. The three signs on the overhead catenary pole are, from top to bottom, a permanent 50 km/h speed limit, a 25 km/h limit on the line to the left as indicated by the arrow (in this case the Beckenham Junction line) and a temporary 25 km/h limit (shown in red) on the Elmers End line. Where temporary speed limit (TSL) signs are displayed, they will continue to apply until the tram passes a red letter T on a white diamond to indicate the end of the TSL. The vertical bar on a white diamond on the right indicates a specific instruction or warning as shown on the sign below, in this case "Ped X" to warn trams that they are approaching a pedestrian crossing. *Alan Yearsley (2)*

CROYDON: TRAM TO TRAMLINK

Above: Trams stabled at Sandilands during Phase 1 of the 2009 blockade on 28 July that year, when the line was closed between East Croydon and Sandilands. Car 2551 is seen on the left, 2543 on the right and 2545 just visible in the background. *Keith Fender*

During the long period of closures in July and August 2009, some of the trams used on the eastern side of the blockade in Phase 1 of the works were swapped by low loader with some of those used on the western side at the beginning of Phase 2 so that as many of them as possible kept to their regular maintenance schedule at the depot. Mobile maintenance teams from the depot carried out light work on the other cars marooned east of the blockade.

When the entire loop was closed, trams running to and from Wimbledon had to terminate at Reeves Corner stop which is normally only served in the eastbound direction as mentioned in chapter 3. During Phase 3 of the 2009 closure from 11 to 31 August a 23 m prefabricated temporary stop, supplied by Cambrian Co Ltd at a cost of £46 000, was established at Dingwall Road, between East Croydon and the junction of George Street and Wellesley Road. This enabled trams to and from Wimbledon to get nearer to East Croydon station than George Street and Wellesley Road stops. The temporary stop was retained by TfL after the blockade for possible use during future engineering closures, and was planned as part of the overall project but not mentioned in TfL publicity about the blockade issued beforehand. Instead the maps on the closure notices suggested that trams would terminate at Wellesley Road and start back from George Street.

SIGNALLING AND SIGNS

Most of the London Trams network uses "Line of Sight" signalling principles as used on other UK light rail systems with street running, allied to a Tram Management System supervisory system. This means that drivers should maintain a speed such that they can stop short of any obstructions simply by using the normal service brakes. Because of this the maximum speed on the on-street sections is generally the same as for motor vehicles on the stretch of road in question. This is usually 50 km/h or 30 mph in built-up areas. Some signalling is still needed at road junctions and crossings and on the various single line sections, however. To avoid confusion with conventional traffic light signals, special white lights are used for the trams.

It should be noted that London Trams, like other TfL modes (i.e. London Underground and the Docklands Light Railway), officially works in metric rather than imperial units, meaning that distances are measured in kilometres and speeds in km/h even though the road signs for motor vehicles on the on-street sections are in miles per hour. Tram speed limits are shown on white diamonds whereas those for other road vehicles are the standard white circles with a red border. As mentioned in chapter 6 the maximum speed on the off-street sections was originally 80 km/h but was reduced to 70 km/h after the Sandilands derailment of 2016 with a new 60 km/h limit also being introduced on the approach to the curve at Sandilands from the tunnel.

THE POWER SYSTEM AND OVERHEAD LINES

Trams operate using a single-arm pantograph to collect current from 750 V DC overhead power lines, this power being fed through a number of substations located across the network. 750 V DC is the standard voltage for almost all UK tramways, except that Blackpool is 600 V DC. The overhead wires are supported by catenary suspension on most of the network, with the on-street sections in Croydon town centre using a mixture of catenary suspension and direct fixings to adjacent buildings.

THE DEPOT AND CONTROL CENTRE

The entire tram fleet is stabled and maintained at Therapia Lane depot adjacent to its namesake stop on the Wimbledon line. By coincidence, this is also close to the site of a former tram depot. There are two tracks inside the depot: one for wheel turning which is done using an Atlas wheel lathe, and one for routine maintenance which is equipped with an inspection pit and synchronised jacks that are used to lift trams up to enable their bogies to be replaced, and there is a washing bay

Chapter 5: The Depot and Infrastructure

Left: The signal for Platform 10b at Wimbledon, with car 2546 in Platform 10a. It is showing five white lights in a horizontal bar, meaning that trams in Platform 10b must wait, and will remain on "stop" until the tram in Platform 10a has departed and cleared the first section. A vertical bar indicates that trams on that track may proceed, a south-east to north-west diagonal bar means that left turning trams may proceed, a south-west to north-east diagonal bar means that trams turning right may proceed, and five white lights in a cross means proceed with caution (equivalent to an amber road traffic light signal). *Robert Pritchard*

Right: CR4000 2531 arrives at East Croydon on 4 July 2021. On the left is a 25 km/h speed limit sign for eastbound trams, and on the right is a give way sign with an instruction to trams on the rarely used track through Platform 2 to give way to trams from the left, affixed to a CCTV camera pole. *Keith Fender*

Above: CR4000 trams 2553 (left) and 2534 (right) undergoing maintenance at Therapia Lane depot on 15 February 2012. *Tony Christie*

CROYDON: TRAM TO TRAMLINK

Above: CR4000 2548 (left) and Variotram 2554 (right) outside Therapia Lane depot on 15 February 2012. *Keith Fender*

Above: The exterior of Therapia Lane depot on 15 February 2012. *Tony Christie*

Chapter 5: The Depot and Infrastructure

Above: The washplant at Therapia Lane depot. *Tony Christie*

Below: The "crab" at Therapia Lane depot. In front of it is the bogie turntable on which tram bogies are turned after being removed and before being put back on the tram. *Keith Fender*

77

CROYDON: TRAM TO TRAMLINK

outside the depot.

Up to four trams, including one on the wheel lathe, can be accommodated inside the depot at a time, with the rest of the fleet being stabled in the yard outside. This has not led to any issues with vandalism or weather damage, as it is a secure site and the vehicles are designed to cope with all weather like their counterparts in Germany and Norway. As well as the trams themselves, also based at the depot is a small yellow battery powered mini-locomotive known as the crab, which is used to shunt trams within the depot site.

Also located at the depot is the control centre where controllers monitor the movement of trams using real-time route diagrams and can view colour CCTV footage from all tram stops and make live announcements either across the entire network or to a specific stop

Above: A CR4000 tram bogie awaits overhaul at Therapia Lane depot. *Keith Fender*

or stops. They can also contact tram drivers and other staff such as roving ticket inspectors by radio, and have direct lines to the emergency services and Network Rail. Calls by passengers from the help points at tram stops go through to one of the duty network controllers.

THERAPIA LANE DEPOT GOES GREENER

In late March 2023, TfL announced that it had received a Government grant of £592 000 to make Therapia Lane depot more environmentally sustainable. The grant was awarded as part of the Department for Energy Security and Net Zero Public Sector Decarbonisation Scheme set up to provide grants for public sector bodies in England to fund heat decarbonisation and energy efficiency measures. This grant will part-fund a number of carbon reduction measures at the depot, with the remaining funding coming from TfL's own decarbonisation investment which forms part of TfL's current Business Plan.

This project will see a mixture of new efficient heat pumps and infrared panel heaters installed to replace ageing and inefficient gas boilers by 2025, thereby reducing dependence on fossil fuels and making significant savings in running costs. The additional electricity consumption from the new heating system is planned to be offset by using 1800 m^2 of south-facing roof space for solar panels along with other energy efficiency measures such as improved insulation and LED lighting. This upgrade forms part of TfL's wider work to decarbonise London's transport network. As one of the UK's largest consumers of electricity, TfL has recently started tendering for a Power Purchase Agreement that will be the first step towards achieving the Mayor of London's target of a net-zero London by 2030. TfL is also actively seeking new locations to plant trees and install other forms of green infrastructure, and is maintaining and developing urban environmentally friendly features, especially in outer London where there is more greenery, which can reduce flooding and support better drainage.

Above: The control centre at Therapia Lane depot on 15 February 2012. The controllers are monitoring the movement of trams using real-time route diagrams. Above their computer monitors are the CCTV screens on which they can view footage from all tram stops. On the right is the engineering works notice board on which details such as name of contractor, time work started, location of worksite and details of work being carried out are entered. *Tony Christie*

CHAPTER 6:
SERVICES

Above: Car 2549 approaches Harrington Road with a Beckenham Junction–Wimbledon service on 17 April 2022. *Robert Pritchard*

There are basically four separate routes, although these overlap on parts of the network particularly in Croydon town centre:

Route 1:	Elmers End–Croydon (early morning and late evening only)
Route 2:	Beckenham Junction–Wimbledon
Route 3:	New Addington–Croydon (certain late evening journeys continue to Therapia Lane and two early morning journeys continue to Wimbledon)
Route 4:	Elmers End–Wimbledon (no early morning or late evening service).

All trams on Route 1 and most on Route 3 take a full circuit of the Croydon town centre loop and then return to their point of origin (Elmers End and New Addington respectively). Driver changes normally take place at East Croydon or Therapia Lane. The current service pattern dates from 25 February 2018 when the routes were reorganised to improve reliability. From that date trams ceased to display their route numbers on their destination indicators.

In the early years of Tramlink, a full timetable for the entire system was displayed at every stop. However, today only the times of first and last trams and service intervals from the stop in question are given, which is not much use if the service frequency is shown as, for example, "every 5–15 minutes". The TfL website also now only features a journey planner. However, the full Tramlink schedules can still be found on the unofficial London Bus Routes website: *www.londonbusroutes.net*

Although there are nominally four different routes in the current service pattern, the TfL Tramlink map shows Routes 1, 2 and 4 as one route in light green and Route 3 as a separate route in dark green. For the first six years of operation there were three routes, each of which were colour coded on official TfL maps: Route 1 (yellow) Elmers End–Wimbledon, Route 2 (red) Beckenham Junction–Croydon, and Route 3 (green) New Addington–Croydon. This pattern lasted until 23 July 2006 when the first reorganisation of routes took place, with Route 1 running from Elmers End to Croydon, Route 2 Beckenham Junction–Croydon and Route 3 New Addington–Wimbledon. On 25 June 2012 a new Route 4 service from Elmers End to Therapia Lane was launched, then from 4 April 2016 this was extended to Wimbledon where the number of trams per hour was increased from eight to 12. The 2018 reorganisation means that shift workers on the industrial estates served by the Wimbledon branch who live along the New Addington line now usually have to change trams in central Croydon, hence the decision to extend certain journeys on Route 3 to or from Wimbledon.

The basic service pattern on each route at the time of writing is:

MONDAYS–FRIDAYS:

Routes 1 and 4:
- Elmers End–Church Street every 15 minutes 04.55–05.55 and 20.10–00.10 westbound, 05.12–06.12 and 20.27–00.27 eastbound.
- Elmers End–Wimbledon every 10 minutes 06.07–19.07 westbound, 06.36–19.56 eastbound.
- Therapia Lane–Elmers End: Eight early morning journeys 04.21–06.42.
- Elmers End–Therapia Lane every 15–30 minutes 19.00–01.00.

CROYDON: TRAM TO TRAMLINK

Above: Car 2553 passes the junction of Dingwall Road just west of East Croydon with a Beckenham Junction service on 30 June 2018. *Tony Christie*

Above: Car 2558 approaches East Croydon with an Elmers End service on 11 August 2018. *Robert Pritchard*

Chapter 6: Services

Right: Car 2536 passes the site of the 2016 Sandilands derailment with a Beckenham Junction or Elmers End to Wimbledon service on 14 July 2021. Note the red poster above the doors and windows reminding passengers that they must wear a face covering. *Keith Fender*

- Beddington Lane–Wimbledon: One early morning journey at 06.24.

Route 2:

- Beckenham Junction–Wimbledon: Every 15 minutes 05.25–05.53 and 19.55–23.40 westbound, 05.26–06.11 and 20.41–00.11 eastbound. Every 10 minutes 05.53–19.43 westbound, 06.11–20.41 eastbound.
- Beddington Lane–Wimbledon: Four early morning journeys at 05.00, 05.15, 06.08 and 06.20 (plus a Route 4 journey at 06.24 as mentioned above).
- Beckenham Junction–Therapia Lane: Every 15 minutes 23.55–01.10.
- Wimbledon–Therapia Lane: Two late evening journeys at 00.26 and 00.41.

Below: Car 2538 traverses the scenic rural section of the New Addington route between Lloyd Park and Coombe Lane stops with a Croydon–New Addington service on 17 April 2022. Since the 2018 service reorganisation most New Addington line services have operated only between New Addington and central Croydon, apart from two early morning westbound trams that are extended to Wimbledon and a number of late evening journeys that terminate at Therapia Lane. *Robert Pritchard*

CROYDON: TRAM TO TRAMLINK

Above: Car 2539 is seen just after leaving Lloyd Park with a New Addington service on 15 April 2022. *Robert Pritchard*

Above: Car 2552 passes through South Norwood Country Park as it heads towards Arena with a Beckenham Junction–Wimbledon service on 17 April 2022. *Robert Pritchard*

Chapter 6: Services

Above: Car 2549 arrives at Platform 10b at Wimbledon with a terminating service (left) while on the right car 2546 waits to depart from Platform 10a on an Elmers End service on 11 June 2023. *Robert Pritchard*

Above: Car 2559 departs from Platform 10b at Wimbledon (right) bound for Beckenham Junction while on the left a cleaner carrying a refuse sack and a litter picker attends to sister Variotram 2564 as it waits to depart from Platform 10a on an Elmers End service on 11 June 2023. *Robert Pritchard*

CROYDON: TRAM TO TRAMLINK

Route 3:
- New Addington–Wimbledon: Two westbound services (04.56 and 05.11)
- New Addington–Church Street: Every 15 minutes 05.25–05.55 and 19.55–00.10 westbound, 05.50–06.20 and 20.20–00.35 eastbound. Every 10 minutes 05.55–06.25 westbound. Every 7–8 minutes 06.32–19.55 westbound, 06.50–20.20 eastbound.
- New Addington–Therapia Lane: Every 15 minutes 20.01–00.01, 00.25.
- Therapia Lane–New Addington: Every 15 minutes 04.14–05.14, 05.25, 05.35, 05.50, 06.05, 06.19.

SATURDAYS:

Routes 1 and 4:
- Elmers End–Church Street every 15 minutes 05.25–07.25 and 19.10–00.10 westbound, 05.42–07.42 and 19.27–00.27 eastbound.
- Beddington Lane–Wimbledon: Three early morning journeys at 07.40, 07.53 and 08.03.
- Elmers End–Wimbledon every 10 minutes 07.40–18.07 westbound, 08.06–18.56 eastbound.
- Therapia Lane–Elmers End: Seven early morning journeys at 04.51, 05.06, 05.21, 07.32, 07.51, 08.02 and 08.12.
- Elmers End–Therapia Lane: Every 10 minutes 18.17–18.57, then 19.17 and 19.47.

Route 2:
- Beckenham Junction–Wimbledon: Every 15 minutes 05.25–07.40 and 18.55–23.40, every 10 minutes 07.53–18.43.
- Wimbledon–Beckenham Junction: Every 15 minutes 05.56–07.41 and 19.41–00.11, every 10 minutes 07.41–19.41.
- Beddington Lane–Wimbledon: Two early morning journeys at 07.35 and 08.18 (in addition to those on Route 4 mentioned above).
- Beckenham Junction–Therapia Lane: Every 15 minutes 23.55–01.10.
- Therapia Lane–Beckenham Junction: Every 15 minutes 04.42–05.57, then 07.17 and 07.47.

Route 3:
- New Addington–Wimbledon: Two westbound services at 04.56 and 05.11.
- New Addington–Church Street: Every 15 minutes 05.25–07.25 and 18.55–00.10, every 10 minutes 07.25–07.55, every 7–8 minutes 07.55–18.55.
- Church Street–New Addington: Every 15 minutes 05.50–07.50 and 19.20–00.35, every 10 minutes 07.50–08.20, every 7–8 minutes 08.20–19.20.
- Therapia Lane–New Addington: Every 15 minutes 04.14–05.29, 06.54, 07.20, 07.35 and 07.49.
- New Addington–Therapia Lane: Every 15 minutes 19.01–19.46 and 00.25–01.10.

SUNDAYS:

Routes 1 and 4:
- Elmers End–Church Street every 15 minutes 06.55–09.40 and 17.55–23.55 westbound, 07.12–09.57 and 18.12–00.12 eastbound.
- Beddington Lane–Wimbledon (westbound only): Every 15 minutes 09.22–10.07.

Above: Some early morning and late evening services start or terminate at Therapia Lane to avoid dead mileage. On 15 April 2022 car 2533 calls at Therapia Lane with a Wimbledon service. *Robert Pritchard*

- Elmers End–Wimbledon: Every 15 minutes 09.56–17.26 westbound, 09.49–17.34 eastbound.
- Elmers End–Therapia Lane: 17.40, 00.10, 00.25, 00.40.
- Wimbledon–Therapia Lane: 17.49, 18.04, 18.19.

Route 2:
- Beckenham Junction–Wimbledon: Every 15 minutes 07.10–23.25 westbound, 07.11–23.56 eastbound.
- Beddington Lane–Wimbledon: Two early morning journeys at 06.45 and 07.00 (plus those on Route 4).
- Beckenham Junction–Therapia Lane: Every 15 minutes 23.40–00.55.
- Wimbledon–Therapia Lane: Two late evening journeys at 00.11 and 00.26.

Route 3:
- New Addington–Church Street: Every 15 minutes 07.10–09.40 and 17.39–23.40, every 7–8 minutes 09.40–17.39.
- Church Street–New Addington: Every 15 minutes 07.35–10.05 and 18.04–00.05, every 7–8 minutes 10.20–18.04.
- Therapia Lane–New Addington: Every 15 minutes 05.59–07.14 and 09.06–09.51.
- New Addington–Wimbledon: Two early morning journeys at 06.41 and 06.56.
- New Addington–Therapia Lane: Every 15 minutes 17.46–18.33 and 23.55–00.40.

Thus, most trams cover the entirety of the route that they serve. However, particularly in the early morning and late evening some journeys start or terminate short at (or are extended to) Therapia Lane (or Beddington Lane in the case of some services terminating at Wimbledon) to ensure that all diagrams start and finish at the depot without the need for "dead mileage" (i.e. empty running). As the first trams start their journeys shortly after 04.00 and the last trams do not complete their journeys until after 01.00 London Trams is almost a 24-hour system. A 30-minute service normally runs through the night on New Year's Eve on the Wimbledon to Beckenham Junction and New Addington routes. The current timetable requires 28 trams in peak hours out of the serviceable fleet of 35 (excluding long-term out of traffic 2551).

All trams normally call at all stops on the route (or section of route in the case of short workings) that they cover, unless instructed to skip a particular stop by control because of exceptional circumstances such as when a stop is closed for repairs or rebuilding or because of overcrowding at a stop caused by a major event taking place nearby. The only exception to this is that Reeves Corner tram stop is served only by eastbound trams with westbound trams passing non-stop. All stops were designated request stops when Tramlink first opened, with the CR4000 trams having originally been equipped with bell pushes and "tram stopping" indicators that were illuminated when a passenger pressed the bell. However, in practice it was rare for a tram to skip a request stop, and by the time the CR4000 fleet received its second refurbishment all stops had officially become compulsory stops so the bell pushes and "tram stopping" signs were then removed from the CR4000s as mentioned in chapter 4, although oddly they do still have signs surrounding the opening button on the inside of each passenger door instructing passengers to press the button to request the next stop or to open the doors. The Variotrams were built with bell pushes and "tram stopping" signs, which remain in situ on these vehicles and still work even though they are not needed.

TICKETING AND FARES

For the first 18 years of Tramlink operation there were ticket machines at all stops supplied by Schlumberger. These accepted coins and notes and gave change, and were operated by turning a dial and then pressing the button in the middle of the dial for the desired fare zone or destination and type of ticket. Wimbledon station also had a simple machine supplied by Metric Accent on the platform that only issued single tickets and accepted coins only with no change given. This was to enable passengers who had transferred from National Rail or London Underground services but who did not already hold a tram ticket to save time. Single tickets bought from ticket machines were valid for 90 minutes from the time of issue.

When Tramlink was being planned, it was envisaged that the network would use point-to-point fares based on distance travelled. However, by the time the system opened zonal fares were the norm

Below: A Tramlink ticket purchased from a machine at West Croydon in 2009. *Keith Fender*

Above: A covered up ticket machine at Lebanon Road in August 2018, shortly after the machines were taken out of use. The note on the cover advises passengers that London Trams are cashless and paper tickets are no longer available from this machine, and gives the address of the TfL webpage containing information on how to pay. *Keith Fender*

Above: The two card readers at Birkbeck. *Alan Yearsley*

CROYDON: TRAM TO TRAMLINK

on most other modes of public transport in Greater London (apart from National Rail services at that time). Because of this, Tramlink started with point-to-point fares, with passengers who were buying ordinary single tickets having to select their actual destination on the machine, but in practice there were only two single fares: 90p from any tram stop east of Wandle Park to anywhere to the east or to the west as far as the Zone 3–Zone 4 boundary at Merton Park, or for journeys west of Wandle Park up to and including Merton Park, while journeys that crossed the zonal boundary at Merton Park were £1.30. From 19 August 2001 zonal tickets replaced point-to-point fares, and the ticket machines were altered so that passengers had only to select the fare zone required rather than their specific stop. From that date, Zone 3 & 4 tickets at £1.30 covered any journey that crossed the zonal boundary while journeys within Zone 4 or between any stop on the Wimbledon line and Wandle Park were priced at 90p. The next major change took place on 4 January 2004 with the introduction of a £1 flat fare across the entire system, thereby bringing Tramlink into line with buses.

In 2018 the increasing move away from paper tickets in favour of Oyster smartcards and contactless payment across all modes of transport in Greater London led to a decision to withdraw all tram stop ticket machines, with paper single tickets ceasing to be available from 16 July of that year. For a few months after that date, the ticket machines had dark blue covers placed over them pending their removal. This move effectively brought London Trams into line with London buses, where cash has no longer been accepted as a form of payment since 2014. In 2017 TfL reported that the proportion of single journeys being paid for at tram stop ticket machines had dropped to 0.3%, 13 years after Oyster card payment was introduced on the trams in 2004 and three years after TfL started to accept contactless payment across all modes in 2014 (this option having been available on buses since 2012).

Instead all tram stops now have card readers, where passengers must validate their Oyster card or contactless debit or credit card or payment device before boarding the tram. Some of the busier stops have four card readers (two per platform), while the quieter stops generally just have one reader per platform. Oyster cards can be topped up online via the TfL website, at any National Rail, London Underground or Docklands Light Railway ticket machine, at newsagents and convenience stores displaying the Oyster Ticket Stop sign and at the London Trams travel information centre at Unit 5, Suffolk House, on the south side of George Street just west of East Croydon. This information centre is open Monday–Friday 09.00–17.00 and also sells Oyster cards, provides travel information and deals with lost property enquiries. Oyster cards can also be bought from Oyster Ticket Stop shops, the larger London Underground ticket machines, and TfL's Visitor Centres at Heathrow Terminals 2 & 3, King's Cross St Pancras, Liverpool Street and Piccadilly Circus Underground stations and opposite Platform 8 at Victoria National Rail station. An initial £7 non-refundable activation fee is payable when buying an Oyster card, and this will be automatically added to the card as credit one year after activation.

The fare structure on London Trams is the same as for TfL buses, meaning that a flat fare (£1.75 in 2023) is charged for each journey. This means that it is only necessary to tap your Oyster or contactless card or device on the card reader before boarding, not after alighting (unlike on National Rail, London Underground or the Docklands Light Railway). Indeed, tapping your card on the reader after leaving the tram will result in you being charged for another journey! However, the tram terminus at Wimbledon is located inside the ticket gates at the station, meaning that the procedure here is slightly more complicated (see below).

Since 2016 a special Hopper Fare has applied on London buses and trams. This means that any additional journeys made within an hour of touching your card on a tram stop card reader or on a bus will be free. It is still necessary to touch on with your card before starting any additional journeys within the same hour, however.

Passengers who only use London Trams and/or buses on any one day will automatically receive a daily price cap (£5.25 at the time

Above: A National Rail validator at Wimbledon. Passengers changing between tram and National Rail or London Underground must touch their cards or devices on these card readers. *Alan Yearsley*

of writing, meaning that this level will be reached if at least three journeys are made in the same day). Those who also use other modes (i.e. any combination of National Rail, LU, DLR, London Overground and/or Elizabeth Line services) will likewise receive a daily price cap according to the number of zones through which they travel (currently £9.60 for Zones 1–3, £11.70 for Zones 1–4, £13.90 for Zones 1–5 or £14.90 for Zones 1–6). A paper One Day Bus and Tram Pass, currently priced at £6 (75p more than the daily bus and tram only price cap), is also still available from the ticket offices at Wimbledon, East or West Croydon, Mitcham Junction, Elmers End and Beckenham Junction stations, and can also be bought from newsagents where it is issued as a special type of single-use Oyster card that is only valid on the day when it is first used.

Paper One Day Travelcards are also valid on trams and do not need to be validated before boarding; however, at the time of going to press Transport for London was expected to withdraw all paper tickets by 2024. In 2023 a paper off-peak One Day Travelcard costs £15.20 for Zones 1–6 and £16.20 for Zones 1–9. Anytime One Day Travelcards, also valid in the morning peak period, are available for £15.20 (Zones 1–4), £21.50 (Zones 1–6) and £27.20 (Zones 1–9).

The entire London Trams network counts as one fare zone for ticketing purposes. Because of this, and as the adjacent National Rail lines run through Zones 3–6, One Day Travelcards (or period Travelcards loaded onto Oyster cards) that cover at least one of Zones 3, 4, 5 or 6 can be used on trams.

Londoners aged 60 or over who hold a 60+ London Oyster photocard (for those aged 60–65) or Older Person's Freedom Pass (aged 66 or over) can travel on London Trams and buses (and most other modes within Greater London) free of charge after 09.00 (09.30 on National Rail) on Mondays–Fridays and any time at weekends and public holidays. Disabled Londoners who hold a Freedom Pass can use their passes at any time on London Trams, buses, the Underground, the DLR, the London Overground and the Elizabeth Line.

However, English National Concessionary Travel Pass holders from outside Greater London can use their passes on London buses but not on trams or any other modes managed by TfL where the full

Chapter 6: Services

Above: Interchange between National Rail and tram services is also available at Elmers End, where car 2545 is seen arriving at the bay platform with a service from Wimbledon on 15 April 2022. *Robert Pritchard*

fare will still be payable. Holders of any national railcards can buy discounted paper off-peak One Day Travelcards, and holders of 16–25, 26–30, Disabled, HM Forces, Senior or Veterans Railcards can have their railcard discount applied to an Oyster card. This can be done by presenting your Oyster card and railcard to a member of staff at an Underground station who will do this for you at a ticket machine. There is no railcard discount on bus or tram fares but this facility is worth considering if also using other modes, in which case holders of qualifying railcards will also benefit from a lower daily price cap than non-railcard holders.

Mobile ticket inspectors patrol the network carrying portable handheld devices that can read Oyster and contactless payment cards and devices to ensure that they have been validated. An £80 penalty fare (reduced to £40 if paid within 21 days) is charged to any passenger found not to have a valid ticket or validated Oyster or contactless payment card or device. Major revenue protection exercises are carried out from time to time, including measures such as sealing off all the exits from a stop and only allowing passengers to board or alight by one door where ticket checks are carried out.

Using Oyster or Contactless on Trams at Wimbledon

If you are entering Wimbledon station to start a tram journey: you must tap your card on the ticket gate to enter the station (unless the gates have been left open) and then tap again on the reader on the tram platform to tell the system that you are travelling by tram, not by National Rail or LU.

If you are arriving by National Rail or LU and leaving by tram: you must tap your card on one of the National Rail yellow readers on Platforms 1–4 (used by LU District Line trains) or Platform 9 (used by Thameslink's Sutton Loop service) to complete your National Rail or LU journey, then tap again on the tram reader located adjacent to the tram platforms to start your tram journey.

There are also pink card readers adjacent to the buffer stops on Platforms 1–4; however, these are for interchanging between LU and National Rail services and should not be used if changing between National Rail or LU and tram services.

If you are arriving by tram and leaving by National Rail or LU: you must tap your card on one of the National Rail yellow readers on Platform 9 (or, if departing by LU, you should use the yellow readers on Platforms 1–4 if you have not already done so on Platform 9). Again you should not use the pink readers, and do not use the reader on the tram platforms.

If you are arriving by tram and exiting the station: simply tap out at the ticket gate as you leave the station. Do not also tap on the tram platform reader.

If you are arriving by tram using Oyster or contactless and leaving by National Rail using a paper ticket (for example if your journey extends beyond the Greater London boundaries): you don't need to do anything with your Oyster or contactless payment card or device on arrival at Wimbledon.

If you are arriving by National Rail using a paper ticket and leaving by tram using Oyster or contactless: simply tap your card or payment device on the tram validator before boarding the tram. Do not touch the National Rail/LU validators or the pink card readers.

The last two scenarios may also occur at any of the other stations that are served both by London Trams and by National Rail services that run to and from stations outside the London Travelcard area, in which case the same procedure applies (for example if you are travelling from Epsom or Dorking to Mitcham Junction and then changing onto the tram).

To avoid confusion between tram and National Rail validators, the bases of the tram validators are green and those used to start or finish a National Rail or LU journey are dark blue. This also applies at the other stations that are served by both modes (see below).

CROYDON: TRAM TO TRAMLINK

Interchanging at Other Stations served by London Trams and National Rail

There are six other stations served both by London Trams and by National Rail services: Beckenham Junction, Birkbeck, East Croydon, Elmers End, Mitcham Junction and West Croydon. Here the procedure for interchanging between tram and heavy rail is more straightforward:

Beckenham Junction and Birkbeck

At these stations the tram stop is located outside the National Rail station (or on the opposite platform in the case of Birkbeck). If changing from train to tram, tap out on the National Rail validator and tap in on the tram validator. If changing from tram to train, do not tap out on the tram validator: simply tap in on the National Rail validator.

East and West Croydon

The tram stop at both these stations is also separate; however, the National Rail stations here have ticket gates. Because of this, if changing from train to tram you will need to tap out at the gate to exit the station and then tap in on the tram validator. If changing from tram to train, do not touch the tram validator: simply tap in on the ticket gate.

Elmers End and Mitcham Junction

At these stations the tram stop is located inside the National Rail station but there are no ticket gates here unlike at Wimbledon. In both cases there are National Rail validators located at the entrance to the station and tram validators adjacent to the tram platform. The procedure here is the same as at Beckenham Junction and Birkbeck (see above).

RIDERSHIP FIGURES

Statistics from the Department for Transport show that the number of journeys on London Trams enjoyed steady growth in the first 15 years of operation, doubling from 15 million in 2000–01 to 31.2 million in 2013–14. After this date ridership figures dropped slightly for the next few years to 27.2 million in 2019–20. The number of journeys in 2020–21 fell to an all-time low of 11.6 million when usage was severely affected by the Covid-19 pandemic, recovering to 19.1 million in 2021–22:

2000–01:	15.0 million[1]	2008–09:	27.2 million	2016–17:	29.5 million
2001–02:	18.2 million	2009–10:	25.8 million	2017–18:	29.1 million
2002–03:	18.7 million	2010–11:	27.9 million	2018–19:	28.7 million
2003–04:	19.8 million	2011–12:	28.6 million	2019–20:	27.2 million
2004–05:	22.0 million	2012–13:	30.1 million	2020–21:	11.6 million
2005–06:	22.5 million	2013–14:	31.2 million	2021–22:	19.1 million
2006–07:	24.6 million	2014–15:	30.7 million		
2007–08:	27.2 million	2015–16:	27.0 million		

[1] Tramlink opened in May 2000 and thus only carried passengers for 11 months of 2000–01.

USEFUL WEBSITES

London Trams (on TfL website): *www.tfl.gov.uk/trams*
Oyster Fares Central: *https://oysterfares.com*
Unofficial Tramlink website (not updated since 2007): *www.croydon-tramlink.co.uk*
British Trams Online (an enthusiasts' website covering the current light rail and heritage tramway scene): *www.britishtramsonline.co.uk*

ACCIDENTS

Apart from the tragic Sandilands derailment of November 2016 (see below) London Trams has a very good safety record to date. Some of the most notable incidents in the network's history so far include:

On 23 September 2001 an 8-year-old boy tried to skateboard in front of a tram in Church Street. His skateboard hit the kerb, throwing him into the doorway of a McDonalds restaurant while the skateboard rolled back onto the tram track and jammed underneath the tram.

Above: Ridership on London Trams enjoyed steady growth for the first 15 years of operation but then dropped slightly for the next few years and substantially during the Covid-19 pandemic, partially recovering to 19.1 million by 2021–22. On 15 April 2022 Variotram 2563 (left) pauses at Beddington Lane bound for Wimbledon while on the right CR4000 2549 forms a Wimbledon–Beckenham Junction service. *Robert Pritchard*

Chapter 6: Services

Above: On 13 September 2008 tram 2520 hit a cyclist at Morden Hall Park foot crossing on the single track section between Morden Road and Phipps Bridge stops, who later died from his injuries. Here car 2547 comes off the single track section as it arrives at Phipps Bridge with a Wimbledon–New Addington service on 13 January 2012. *Keith Fender*

On 7 September 2002 a tram hit a 78-year-old woman as she crossed over the track at the Croydon end of New Addington tram stop. She was taken to Mayday Hospital with severe head injuries and pronounced dead on arrival.

On 14 December 2002 tram 2534 hit and killed a 34-year-old woman near Lloyd Park tram stop as it travelled towards Croydon, despite the driver having applied the hazard brakes on noticing the woman.

On 22 June 2003 tram 2531 hit and killed a 57-year-old man on the foot crossing at the junction of Addiscombe Road and Chepstow Road, between Lebanon Road and Sandilands tram stops. It was thought that the man had crossed behind an eastbound tram when the pedestrian signal was showing a red man, and had then stepped into the path of car 2531 which was heading towards Croydon. Tram drivers are trained to ring their bells as they pass other trams, but this warning did not work on this occasion.

On 21 October 2005 tram 2530 derailed as it passed over the points from the single to the double track section at Phipps Bridge when travelling from Wimbledon towards Croydon. A similar incident occurred here with tram 2532 on 26 May 2006, also on an eastbound service. In both cases the RAIB found that the points had failed to return to normal after the previous tram had passed. In the case of the May 2006 derailment it was also found that the points had changed position as the tram passed over them, and that the indicator showing which way the points were set had suffered from poor visibility.

On 7 September 2008 a man was killed after a bus collided with tram 2534 in George Street in central Croydon. Six other people were injured and had to be taken to hospital. The bus driver was convicted of causing death by dangerous driving in December 2009 and sentenced to four years in prison.

On 13 September 2008 tram 2530 hit a cyclist at Morden Hall Park foot crossing between Morden Road and Phipps Bridge stops. The cyclist was seriously injured and later died. It was found that the immediate cause of this accident was that the cyclist rode onto the crossing without looking at the tram, and it is thought that he may have been wearing headphones, which would have prevented him from hearing any audible warnings.

On 5 April 2011 a woman tripped over and was dragged under a moving tram at East Croydon. She was then taken to hospital in a serious condition. It is thought that she was running to catch the tram.

On 17 February 2012 a westbound tram derailed after passing over facing points as it approached East Croydon and headed for Platform 3. The points moved under the leading bogie, forcing the centre and rear bogies towards Platform 2, the opposite side of the island platform. The centre bogie then derailed. Around 100 passengers were detrained but there were no reported injuries. The RAIB reported that a track circuit had failed to lock the points in response to the approaching tram.

On 16 May 2012 a woman was hit by a tram and seriously injured as she crossed the track on a foot crossing near Sandilands tram stop. She then fell into the space between the platform and the tram and was trapped. The RAIB found that the tram driver did not apply the hazard brake after hitting the woman, and recommended that drivers be instructed to do so after striking a pedestrian.

On 13 April 2013 a Beckenham Junction-bound tram departed from Lebanon Road and Sandilands tram stops with all of its doors open on the left-hand side. Some of the doors closed automatically while the tram was moving, but one set of doors stayed open. A controller monitoring the tram on CCTV noticed that the doors were open on leaving Sandilands, and arranged for the tram to be stopped. There were no injuries. An RAIB investigation found that the driver had been trying to resolve a fault with the tram and had inadvertently operated a fault override switch that disables safety systems such as the door-traction interlock.

CROYDON: TRAM TO TRAMLINK

On 7 February 2016 five people were injured when a car hit tram 2535, which was going round a curve near Wellesley Road and was derailed as a result.

THE 2016 SANDILANDS ACCIDENT

By far the most serious accident in the history of Tramlink/London Trams to date was the derailment at Sandilands of 9 November 2016. At around 06.07 CR4000 tram 2551, forming a service from New Addington to Wimbledon, derailed and overturned as it negotiated the sharp left turn out of Sandilands Tunnel towards Sandilands tram stop. The tram overturned on its right side and came to a stand some 27 yards (25 m) beyond the point of derailment. Around 34 passengers were ejected through the broken windows with seven being killed and 62 injured, 19 of them with serious or life-changing injuries. Only one person was unhurt. Following the incident tram driver Alfred Dorris was arrested by the British Transport Police on suspicion of manslaughter, and was released on bail until May 2017.

This was the first fatality on a UK tramway since 1959 when a tram in Glasgow collided with a lorry and caught fire, killing the driver and two passengers. It was also the worst UK tram accident since 1917 when a Dover Corporation tram ran down Crabble Hill, killing 11 people and injuring 60. In terms of the number of people killed, the Sandilands derailment was the worst fatality on any UK rail network since the Great Heck crash on the East Coast Main Line on 28 February 2001 in which ten people died and 82 were seriously injured. Since then there had been only two other incidents involving seven fatalities: the Potters Bar derailment of 10 May 2002 and the Ufton Nervet level crossing crash on 6 November 2004.

An interim Rail Accident Investigation Branch report into the incident published a week later found no evidence of any track faults or obstructions on the track that could have contributed to the derailment. A second RAIB interim report published on 20 February 2017 concluded that driver had "lost awareness" prior to the tram derailing, and that there was evidence suggesting that he had made a late brake application as he entered the sharp curve at 46 mph (73 km/h). It was initially thought that he had approached the curve, which has a 13 mph (20 km/h) speed limit, at 43.5 mph.

After the derailment, trams ran only between Wimbledon and East Croydon, New Addington– Addington Village, and Beckenham Junction–Harrington Road with services being suspended on the Elmers End branch and the rest of the Beckenham Junction and New Addington lines. The damaged tram was removed from the scene of the accident on the morning of 12 November with services across the entire network resuming on 18 November. Following the publication of the first interim RAIB report, before the full service resumed chevron warning signs were installed on the approaches to Sandilands Junction and the 20 km/h speed limit brought closer to the exit from the tunnel. Similar measures were implemented at other sharp curves across the Tramlink network: in the opposite direction between Sandilands and Lloyd Park to the south of Sandilands Tunnel, on the approach to Sandilands from the Beckenham Junction/Elmers End branch and on the bend between Birkbeck and Harrington Road. A new 60 km/h limit was also introduced between the exit from Sandilands Tunnel and the start of the 20 km/h limit. Across the rest of the Tramlink network, the maximum speed was reduced from 80 km/h to 70 km/h.

In March 2017 it was reported that both TCL and TOL had admitted liability for the incident. This meant that survivors and the families of victims would not have to sue for compensation. The final RAIB report into the accident, published in December 2017, suggested that the driver had had a microsleep seconds before the tram overturned. This report found that the speed limit sign for the curve at Sandilands was not visible for drivers until they had passed the point where a brake application was needed to reach the permitted speed. Instead, drivers were expected to know from their route knowledge that this was necessary; however, in this instance the driver was thought to have suffered a lapse in concentration as he approached the curve. It was suggested that this may have led him to forget where he was until he saw the speed limit sign, which is also not always clearly visible in hours of darkness such as on this occasion. Recommendations made in the report were:

- The Office of Rail and Road (ORR) to work with the UK tram industry to set up a body to promote more effective UK-wide co-operation on safety matters and developing good practice guidance and common standards.
- All UK tram operators to review operational risks and safety measures associated with the design, operation and maintenance of tramways.
- UK tram operators to work together to develop speed control systems to reduce the speed of trams automatically if they approach high risk locations at speeds that could cause the tram to derail or overturn.

Below: A map showing the location of the accident. *Courtesy RAIB*

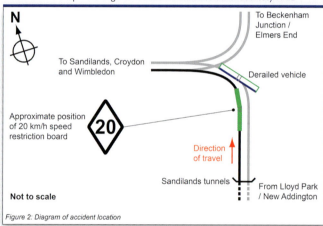

Figure 2: Diagram of accident location

Right: The derailed car 2551 after coming to rest on its side on the curve approaching Sandilands stop. *Courtesy RAIB*

Chapter 6: Services

Right: Gavin Barwell, the then MP for Croydon Central, talks to the BBC near the site of the derailment on the evening of the incident, 9 November 2016. *Keith Fender*

- UK tram operators to develop technology to monitor a tram driver's attention levels and initiate appropriate automatic responses where a low level of alertness was identified.
- Tram operators to review signage, lighting and other visual information cues on segregated and off-street sections of track at high risk locations such as on the approaches to tight curves, in line with the cues provided for road users that are designed according to current UK highway standards.
- Windows and doors to be strengthened to improve the containment of passengers in an accident.
- Tram operators to make emergency lighting on trams more resilient so that it cannot be unintentionally switched off or disconnected in an emergency.
- Emergency exits from a derailed tram to be reviewed.
- ORR to review the regulatory framework for tramways and its long-term strategy for supervision of the sector.
- TOL and London Trams to commission an independent review of its process for assessing the risks associated with the operation of trams. This recommendation was made because the RAIB report found a lack of awareness and understanding among the management of Tramlink about the risk of a tram overturning.
- TOL to review and improve its management of fatigue risk affecting drivers with reference to the ORR's good practice guide.

These two recommendations were made because of an incident just nine days before the derailment in which a tram approached the curve at Sandilands too fast and came close to derailing but was reported by a member of the public, not by the driver:

- TOL to undertake a review covering the way that it learns from operational experience, including creating a "just culture" where staff are more likely to report incidents and safety concerns.
- Management systems to be improved to better follow up public and employee comments where these relate to safety risks.
- On-tram CCTV equipment to be better maintained and inspected to ensure that CCTV images are always available to investigate incidents. The RAIB found that the on-board CCTV equipment on tram 2551 was not working at the time of the derailment.
- TOL to update tram maintenance manuals to take account of experience gained and modifications made since the trams entered service. Investigators found that some of the London Trams maintenance instructions were out of date.

Above: Variotram 2565 heads towards Sandilands tram stop as it passes the site of the 2016 derailment with a Beckenham Junction or Elmers End to Wimbledon service on 25 January 2021. On the left can be seen the chevron warning sign installed after the crash. *Keith Fender*

CROYDON: TRAM TO TRAMLINK

Above: The memorial to the 2016 derailment, bearing the names of the seven passengers killed, at the junction of Sandilands and Addiscombe Road, close to the site of the accident. In the background is the overbridge that carries Sandilands and Addiscombe Road over the tramway towards Beckenham Junction and Elmers End. *Keith Fender*

In terms of strengthening the tram doors and windows, the RAIB investigation found that the windows were made of toughened glass, in line with regulations applying to trams and buses, and were not therefore strong enough to prevent passengers being ejected from the vehicle. Because of this, the RAIB recommended that laminated glass should be used on trams, as this would have held the passengers inside the tram as it derailed and had already been made mandatory for conventional heavy rail vehicles after a number of accidents in which passengers had been ejected from trains.

In response to recommendation 4, by October 2017 TfL had installed a Driver Protection Device on all trams. These devices, manufactured by Seeing Machines, shine an infrared light into the driver's face and can generate an alert and vibrate the driver's seat if head or eye movements indicate a lack of attention from the driver. This prompted two 24-hour strikes by tram drivers on 13 November and 6 December that year, as drivers' union ASLEF claimed that these devices posed a number of health and safety issues including the risk of headaches, dry eyes, blurred vision and potential eye damage.

By the end of 2019, a new on-board automatic braking system had been fitted to all trams. This system consists of yellow beacons installed on the track that measure the speed of a passing tram, and if the tram is found to be exceeding the speed limit its brakes are automatically gradually applied.

Meanwhile, by late October 2019 the Crown Prosecution Service had completed its investigation into the derailment and concluded that there was insufficient evidence to prosecute the driver, TfL or TOL. The CPS stated that although there was evidence of negligence by the driver, it could not be classed as "gross negligence" meaning that none of the parties involved could be charged with manslaughter by gross negligence.

Following this decision, the date for an inquest was set for 19 October 2020 but was postponed because of Covid restrictions, eventually starting on 17 May 2021 when a jury was sworn in. On 22 July the jury returned a verdict of accidental death. This prompted families of the victims to demand a fresh inquest. Although the inquest heard detailed evidence from the RAIB and from British Transport Police investigators, families of the victims believed that senior rail bosses should also be called to give evidence.

In March 2022 the ORR announced that it intended to press charges against TfL and TOL for health and safety failings relating to the derailment. TfL and TOL were both alleged to have failed to ensure the health and safety of passengers on the network as far as reasonably practicable. Both bodies pleaded guilty at a hearing at Croydon Magistrates' Court on 10 June. The tram driver was also charged with failure to take reasonable care of passengers whilst employed as a driver. He pleaded not guilty and was released on unconditional bail pending a further hearing at Croydon Crown Court on 8 July at which judge Mr Justice Fraser applied to transfer Mr Dorris's case for trial at the Old Bailey or Southwark Crown Court. His trial at the Old Bailey started on 16 May 2023, and on 19 June he was found not guilty. On 26 July TfL was fined £10 million and TOL £4 million at a hearing at the Old Bailey.

Memorials

Two memorials to the crash were unveiled in late 2017, one close to the site of the derailment at the junction of Sandilands (the road from which the tram stop gets its name) and the A232 Addiscombe Road, and one in Market Square on Central Parade, New Addington. The memorial at Sandilands bears the names of all seven passengers killed in the accident. On 9 November 2022, the sixth anniversary of the derailment, an open air memorial service was held in New Addington at which flowers were laid by the memorial in honour of the victims.

Similar Incidents

Shortly after the Sandilands derailment there were media reports of a tram having almost derailed at the same location on at least two occasions, on 22 and 31 October 2016. In both cases it was reported that a passenger had felt as if the tram on which they were travelling was going too fast around the curve at Sandilands. Then on the morning of 17 May 2017 a passenger filmed a driver asleep in the cab of a tram. Passengers on the Wimbledon-bound tram involved first noticed that something was wrong when the tram failed to continue its journey after being held at a signal near East Croydon tram stop. Most of the 50 people on board decided to alight and walk to their destinations, one of whom knocked on the cab window to alert the driver who was then suspended pending an investigation.

A month earlier, in April 2017, a BBC investigation for the Victoria Derbyshire programme found that at least three trams had been recorded exceeding the speed limit on the same section of line since the derailment. There were also reports of a number of drivers who admitted having fallen asleep when driving trams and that the driver safety device (a.k.a. "dead man's handle"), which should automatically apply the emergency brake if the driver fails to activate this device every few seconds, had not worked on any of these occasions. There was no suggestion that a problem with the safety device had been responsible for the Sandilands derailment, however.

CHAPTER 7:
FUTURE DEVELOPMENTS

Above: Perhaps the most serious proposal to date has been for an extension to Crystal Palace, which would involve Tramlink taking over the Crystal Palace–Beckenham Junction line. Even without trams, Crystal Palace station is already served by London Overground and Southern trains. On 4 August 2018 London Overground unit 378 151 departs from one of the terminal bays with the 11.02 to Highbury & Islington with the Grade II listed station building providing a magnificent backdrop. *Robert Pritchard*

By the time this book appears, London Trams may be the only UK light rail system never to have been extended beyond its original network. The Edinburgh tramway also fell into that category until the recent opening of the Newhaven extension, and the Supertram network in our home city of Sheffield remains unchanged apart from the tram-train route to Rotherham. At the time of writing this also applied to the Blackpool tramway (at least in terms of the remaining route post-1963 between Blackpool and Fleetwood) but the extension to Blackpool North station was expected to open in the next few months.

A number of extensions to the London Trams network have been mooted over the years but to date none of these schemes have come to fruition, probably largely for financial reasons. Perhaps the most serious proposal was an extension to Crystal Palace, which would have involved taking over the Crystal Palace–Beckenham Junction heavy rail line with

Right: One option for the Tramlink extension to Sutton envisaged the conversion of the Wimbledon–Sutton line to light rail operation. The Baitul Futuh Mosque at Morden South provides an impressive backdrop to 700 023 arriving with the 13.37 St Albans–Sutton on 30 November 2022, but the quiet station feels a little unloved and quite rural, with some rampant vegetation! *Robert Pritchard*

CROYDON: TRAM TO TRAMLINK

two new intermediate stops between Birkbeck and Crystal Palace at Penge Road and Anerley Road. The line would then have continued beyond Crystal Palace station to serve the nearby Crystal Palace Park. TfL considered three different route options between Crystal Palace station and the park: on-street via Anerley Hill, via the existing railway alignment, or via Anerley Road and part of Crystal Palace Park.

This scheme had been fully designed and funded during Labour London Mayor Ken Livingstone's period of office (2000–08) but was shelved by his Conservative successor Boris Johnson despite having promised to take the project forward in the 2008 and 2012 mayoral election campaigns. The scheme also did not feature in Transport for London's four-year plan for public transport in London published in 2011.

The only other extension that TfL has seriously considered to date is a new line from Wimbledon or Colliers Wood to Sutton, known as the Sutton Link. Three possible routes have been considered, all of which would end in a loop around Sutton town centre. Option 1 would start from South Wimbledon Underground station and run via Morden Road (where there would be an interchange with the existing Wimbledon–Croydon line), St Helier Avenue and Rose Hill, while option 2 would start from Colliers Wood Underground station and run via Church Road to the existing Belgrave Walk tram stop, then via Morden Road until it joined St Helier Avenue and would then continue via the same route as option 1. Both these options could be used for a tram or bus rapid transit route, and could include a loop serving St Helier Hospital. Option 3 would have involved taking over the existing Wimbledon–Sutton heavy rail line. In late 2018 a TfL consultation on the scheme was launched, and in February 2020 TfL declared its preference for option 2 involving a tram route between Colliers Wood and Sutton. TfL estimated that such a route would cost £425 million (at 2018 prices) but would be dependent on funding from Merton and Sutton Borough Councils. However, in July 2020 TfL announced that it had shelved plans for an extension to Sutton following London Mayor Sadiq Khan's funding deal with the Government that had been made necessary by the financial crisis faced by TfL following the Covid pandemic.

In 2002 a consortium of councillors, local business leaders and experts in the Royal Borough of Kingston-upon-Thames drew up outline proposals for a tram network in its area known as K-SMART (Kingston, Surbiton, Malden and Richmond Transit). This was initially envisaged as a self-contained network including routes from Kingston to New Malden and Chessington taking over the existing railway line to Chessington South, with a longer-term aspiration to extend the New Malden route to Worcester Park and Sutton where it would have joined with the proposed new Tramlink route. The South London Partnership was then formed in 2004, comprising a number of local councils and business organisations, to promote the case for extensions to Tramlink.

Other extensions that have been considered in the past include Beckenham Road–Lewisham, Beckenham Junction–Bromley, Croydon–Purley (including a further potential extension to Coulsdon), New Addington–Biggin Hill, and a second Croydon town centre loop via Dingwall Road and Lansdowne Road to relieve congestion on the existing central Croydon loop. However, to date these schemes have got no further than feasibility studies. In the case of

the Dingwall Road loop, Croydon Council objected to this scheme in 2015 on the grounds that its vision for Croydon was of a more pedestrian and cycle friendly place, meaning that trams using this loop would have to run at a reduced speed.

In late 2014 TfL unveiled plans for a series of enhancements to the existing network for the period to 2030, entitled Trams 2030. At this time ridership levels were forecast to reach 60 million passenger journeys by 2031, partly as a result of the then planned rebuilding of the Whitgift shopping centre including a merger with the adjacent Centrale retail complex. Although Croydon Council approved the redevelopment plan in November 2013, the plan was later revised twice and thus delayed, with the Covid pandemic causing further uncertainty.

The Dingwall Loop formed part of Trams 2030, with other proposals including doubling the single track flyover at Wandle Park, a new turnback or loop near Reeves Corner, additional stabling facilities at Therapia Lane depot or at Harrington Road or Elmers End to accommodate the extra trams needed to increase service frequencies, a second platform at Elmers End, and lengthening tram stops to accommodate longer trams, either longer vehicles to replace the CR4000s or the operation of pairs of trams coupled together.

In the current economic and financial climate it unfortunately seems unlikely that any extensions or significant upgrades to the current system will see the light of day in the near future apart from replacement of the CR4000 tram fleet, and possibly the second platform at Elmers End which would offer the greatest benefit for a relatively low spend.

THE CROSS-RIVER AND WEST LONDON TRAM SCHEMES

For the sake of completeness, mention should also be made here of two other London tram schemes that were promoted in the early to mid 2000s but ultimately abandoned.

The **Cross-River Tram** scheme envisaged a 10 mile (16 km) north-south tram route from Camden Town via King's Cross, Euston, Holborn and Waterloo to Brixton and Peckham. It was first promoted by TfL in 2006 as a way of relieving overcrowding on the London Underground and regenerating areas poorly served by public transport such as the Aylesbury Estate in Southwark. However, by the autumn of 2007 London Mayor Ken Livingstone had reported that Camden Council had opposed the scheme. Because of this, the Mayor suggested that the southern sections from Brixton and Peckham to Waterloo could be built first with the northern sections from Waterloo to Euston, King's Cross and Camden Town following at a later date. In May 2008 Livingstone's newly elected Conservative successor Boris Johnson said he would review the scheme for which no Government

Right: In 2002 a consortium of councillors, local business leaders and experts in the Borough of Kingston drew up proposals for a tram network in the area that would have taken over the existing Chessington South branch. On 22 October 2022 South Western Railway Class 455 units 5719 and 5854 arrive at Tolworth with a Waterloo–Chessington South service. *David Bosher*

Chapter 7: Future Developments

Left: Another scheme that has been mooted in the past envisaged an extension beyond the New Addington terminus to Biggin Hill. On 5 April 2019 Variotram 2563 calls at King Henry's Drive with a New Addington–West Croydon service. *Robert Pritchard*

funding had yet been secured, and in late 2008 TfL announced that the project had been cancelled. Instead TfL said it would explore alternative solutions such as increased capacity on existing Underground lines.

The **West London Tram** was planned to run along the A4020 Uxbridge Road corridor from Shepherd's Bush and Shepherd's Bush Market stations to Uxbridge via Acton, Ealing, Southall, Hayes and Hillingdon. It was first promoted by Mayor Ken Livingstone in 2002 and would have followed the same route as the original London tram route 7, which ran from 1904 until 1951. This was then replaced by trolleybus route 607, which itself was later superseded by bus routes 207, 427 and 607.

Mayor Ken Livingstone promoted the West London Tram as a solution to traffic congestion in the areas that it would serve. It was planned to run on-street along the entire length of the route, using a mixture of reserved track and stretches shared with motor vehicles.

However, the scheme was a contentious issue in the areas to be served, with several consultation exercises and opinion polls showing narrow majorities both for and against the plan. Opposition was strongest in the Borough of Ealing, where a campaign entitled Save Ealing's Streets was formed in 2004. This group expressed concerns that a number of sections of the Uxbridge Road were not wide enough to accommodate two lanes of traffic and two tram tracks. Because of this they believed that a significant amount of traffic could end up being diverted through residential areas. This group also took the view that the reduction in road traffic would not be as high as TfL had suggested, citing TfL's own impact studies of Croydon's Tramlink system. The group pointed out that Tramlink ran mostly on former railway lines with only a short on-street section, with traffic displacement being managed by the construction of a new bypass. By contrast, the West London Tram was planned to run entirely on a major highway with adjoining narrow residential streets and no scope for new roads or widening existing roads.

The scheme was also opposed by Hillingdon and Hammersmith & Fulham Councils, while Ealing Council initially supported the project but opposed it following a change of control from Labour to Conservative at the 2006 council elections. The project was postponed indefinitely by TfL in August 2007 following the announcement that the Government was to go ahead with the Crossrail scheme. Instead TfL pledged to work with the three borough councils to increase bus provision but also indicated that the tram scheme could be revisited if further public transport capacity were needed after Crossrail (now the Elizabeth Line) had opened.

Campaign group Save Ealing's Streets suggested a monorail as an alternative solution, arguing that such a system would cost less than half as much per kilometre as a tramway because it would not involve digging up the road and diverting the utilities and would also not require road traffic to be diverted. Meanwhile another campaign group entitled Trolley Buses for West London, and a group

ABBREVIATIONS USED IN THIS BOOK

- ABB: Allmänna Svenska Elektriska Aktiebolaget (ASEA) Brown Boveri.
- AGV: Automated Guided Vehicle.
- ASLEF: Associated Society of Locomotive Engineers and Firemen (train drivers' union).
- BETCo: British Electric Traction Company.
- BR: British Rail.
- C&NTCo: Croydon & Norwood Tramways Company.
- CCTV: Closed circuit television.
- CPS: Crown Prosecution Service.
- CTCo: Croydon Tramways Company.
- DC: Direct current.
- DLR: Docklands Light Railway.
- EMU: Electric Multiple Unit.
- EPB: Electro Pneumatic Brake.
- LB&SCR: London, Brighton & South Coast Railway.
- LCC: London County Council.
- LED: Light-Emitting Diode.
- LPTB: London Passenger Transport Board.
- LRO: Light Railway Order.
- LRT: London Regional Transport.
- LSWR: London & South Western Railway.
- LTCo: London Tramways Company.
- LU: London Underground.
- LUT: London United Tramways.
- MET: Metropolitan Electric Tramways.
- N&DTCo: Norwood & District Tramways Company.
- NMTCo: North Metropolitan Tramways Company.
- ORR: Office of Rail and Road.
- RAIB: Rail Accident Investigation Branch.
- SMET: South Metropolitan Electric Tramways & Lighting Company.
- SR: Southern Railway.
- STTCo: Steam Tramways Traction Company.
- TCL: Tramtrack Croydon Ltd.
- TfL: Transport for London.
- TOL: Tram Operations Ltd.
- VEP: Vestibuled Electro-Pneumatic brake unit.
- W&CR: Wimbledon & Croydon Railway.
- WHCT: West Ham Corporation Tramways.

of transport consultants named the Electric Tbus Group, suggested that a trolleybus system would offer a cheaper and more flexible alternative (see our sister publication **Nottingham: Tramway to Express Transit** for a detailed discussion of the advantages and disadvantages of trolleybuses compared to trams).

Also worth a mention is the **Greenwich Waterfront Transit scheme** involving a route from Abbey Wood to North Greenwich Transport Interchange via Thamesmead and Woolwich with a possible extension to Greenwich town centre. Originally envisaged as a tramway when the scheme was first promoted in 1997, by 2001 trolleybuses were the favoured option and this was later changed to conventional diesel buses but using a mixture of bus lanes and a guided busway for part of the route with the potential for conversion to a tramway at a later date. However, the project was cancelled on financial grounds by Mayor Boris Johnson in March 2009. Together with the East London Transit bus rapid transit scheme opened between 2010 and 2013, the Greenwich Waterfront Transit formed part of the Thames Gateway Transit project.

THE JUBILEE LINE EXTENSION: WHAT MIGHT HAVE BEEN

Apart from the never realised extensions described above, another scheme that could have altered parts of the Tramlink network had it been implemented as originally envisaged was the Jubilee Line extension of the London Underground. When the Jubilee Line was originally being planned in the 1960s and 1970s it was to be named the Fleet Line. As well as taking over the existing Baker Street–Stanmore branch of the Bakerloo Line with an extension south from Baker Street to Charing Cross, the line was also expected to continue beneath Fleet Street (hence the name Fleet Line) to Fenchurch Street, from where it would be extended in later stages to Thamesmead via London's Docklands (with the Thamesmead branch being named the River Line) and to Lewisham, eventually taking over the existing lines to Hayes and Addiscombe. The Woodside–Sanderstead line was not included as part of this scheme and was thus presumably expected to close before the full Fleet Line extension had been completed (and did indeed close in 1983).

In the event, the changing political situation and the state of the country's finances meant that the line was never extended beyond Charing Cross until the late 1990s by which time the scheme had evolved into the present-day Jubilee Line extension to Stratford via North Greenwich. Ironically, in more recent times the Bakerloo Line has been proposed for extension to Lewisham, again potentially taking over the Hayes line. This project has now been put on hold in view of TfL's financial situation although in early 2021 the Department for Transport did safeguard the route against any developments that could hinder its construction.

One can only speculate as to how the Jubilee Line extension to Hayes and Addiscombe could have impacted the Tramlink project, though it is unlikely that the Elmers End tram route would then have existed, and the Beckenham Junction line would also either not have been built or would have had to take a completely different and somewhat more circuitous route. For example trams to Beckenham Junction (if this route had still been built) could have turned left off the existing route onto Cherry Orchard Road and continued via the B243 Morland Road to an Underground/tram interchange at Blackhorse Road. Here they would have terminated if the route had not continued all the way to Beckenham Junction, or the tramway could then have run alongside the Jubilee Line (maybe with only one single tram track and one single Underground track depending on space limitations) as far as Arena (where another tram/Tube interchange could have been built) from where they would have followed the existing route.

The New Addington line could either have taken the existing route via Addiscombe Road and Sandilands or run via Cherry Orchard Road and Lower Addiscombe Road past Addiscombe station (which could likewise have been transformed into a tram/Tube interchange), joining the existing alignment at the site of Bingham Road station/Addiscombe tram stop (which might then have been named Bingham Road to avoid confusion with the other Addiscombe station). There is also the question of whether the so far never realised extensions such as Crystal Palace and Sutton might then have been built, as the initial network would have been smaller so at least some of the funds saved in that way could potentially have been used to build those routes.

> **FURTHER READING ON TRAMLINK AND CROYDON'S ORIGINAL TRAMWAYS AND TROLLEYBUSES**
>
> - Croydon Tramlink: A Definitive History, Gareth David, Pen & Sword, 2020.
> - Croydon Tramways, Robert J Harley, Capital Transport Publishing, 2004 (out of print).
> - Croydon's Tramways, John B Gent and John H Meredith, Middleton Press, 1994.
> - Croydon's Trolleybuses, Terry Russell, Middleton Press, 1996.
> - London Transport Tramways 1933–1952, E.R. Oakley and C.E. Holland, London Tramways History Group, 1998 (out of print).

Above: Parts of the Tramlink network, such as the Elmers End branch, might never have been built if the Jubilee Line had taken over the Hayes and Addiscombe lines as envisaged in the 1960s and 1970s. More recently TfL has proposed to extend the Bakerloo Line to Lewisham and eventually to Hayes. On 4 July 2016 a Bakerloo Line train, formed of 1972 Tube Stock with driving car 3267 leading, arrives at the line's current terminus of Elephant & Castle. *Robert Pritchard*